Frontiers in Applied Dynamical Systems:
Reviews and Tutorials

Volume 3

More information about this series at http://www.springer.com/series/13763

Frontiers in Applied Dynamical Systems: Reviews and Tutorials

The Frontiers in Applied Dynamical Systems (FIADS) covers emerging topics and significant developments in the field of applied dynamical systems. It is a collection of invited review articles by leading researchers in dynamical systems, their applications, and related areas. Contributions in this series should be seen as a portal for a broad audience of researchers in dynamical systems at all levels and can serve as advanced teaching aids for graduate students. Each contribution provides an informal outline of a specific area, an interesting application, a recent technique, or a "how-to" for analytical methods and for computational algorithms, and a list of key references. All articles will be refereed.

Editors-in-Chief

Christopher K R T Jones, *The University of North Carolina, North Carolina, USA*

Björn Sandstede, *Brown University, Providence, USA*

Lai-Sang Young, *New York University, New York, USA*

Series Editors

Margaret Beck, *Boston University, Boston, USA*

Henk A. Dijkstra, *Utrecht University, Utrecht, The Netherlands*

Martin Hairer, *University of Warwick, Coventry, UK*

Vadim Kaloshin, *University of Maryland, College Park, USA*

Hiroshi Kokubu, *Kyoto University, Kyoto, Japan*

Rafael de la Llave, *Georgia Institute of Technology, Atlanta, USA*

Peter Mucha, *University of North Carolina, Chapel Hill, USA*

Clarence Rowley, *Princeton University, Princeton, USA*

Jonathan Rubin, *University of Pittsburgh, Pittsburgh, USA*

Tim Sauer, *George Mason University, Fairfax, USA*

James Sneyd, *University of Auckland, Auckland, New Zealand*

Andrew Stuart, *University of Warwick, Coventry, UK*

Edriss Titi, *Texas A&M University, College Station, USA* and *Weizmann Institute of Science, Rehovot, Israel*

Thomas Wanner, *George Mason University, Fairfax, USA*

Martin Wechselberger, *University of Sydney, Sydney, Australia*

Ruth Williams, *University of California, San Diego, USA*

C. Eugene Wayne • Michael I. Weinstein

Dynamics of Partial Differential Equations

Review 1: C. Eugene Wayne: Dynamical Systems and the Two-dimensional Navier-Stokes Equations

Review 2: Michael I. Weinstein: Localized States and Dynamics in the Nonlinear Schrödinger/Gross-Pitaevskii Equation

 Springer

C. Eugene Wayne
Department of Mathematics
and Statistics
Boston University
Boston, MA, USA

Michael I. Weinstein
Department of Applied Physics
and Applied Mathematics
Columbia University
New York, NY, USA

ISSN 2364-4532   ISSN 2364-4931   (electronic)
Frontiers in Applied Dynamical Systems: Reviews and Tutorials
ISBN 978-3-319-19934-4   ISBN 978-3-319-19935-1   (eBook)
DOI 10.1007/978-3-319-19935-1

Library of Congress Control Number: 2015942469

Mathematics Subject Classification (2010): 78A40, 35Q55, 74J30, 37Kxx, 35Q40, 35Q30, 37L10, 37L30, 37L45

Printed on acid-free paper

Springer International Publishing AG Switzerland is part of Springer Science+Business Media (www.springer.com)

Preface to the Series

The subject of dynamical systems has matured over a period more than a century. It began with Poincaré's investigation into the motion of the celestial bodies, and he pioneered a new direction by looking at the equations of motion from a qualitative viewpoint. For different motivation, statistical physics was being developed and had led to the idea of ergodic motion. Together, these presaged an area that was to have significant impact on both pure and applied mathematics. This perspective of dynamical systems was refined and developed in the second half of the twentieth century and now provides a commonly accepted way of channeling mathematical ideas into applications. These applications now reach from biology and social behavior to optics and microphysics.

There is still a lot we do not understand and the mathematical area of dynamical systems remains vibrant. This is particularly true as researchers come to grips with spatially distributed systems and those affected by stochastic effects that interact with complex deterministic dynamics. Much of current progress is being driven by questions that come from the applications of dynamical systems. To truly appreciate and engage in this work then requires us to understand more than just the mathematical theory of the subject. But to invest the time it takes to learn a new sub-area of applied dynamics without a guide is often impossible. This is especially true if the reach of its novelty extends from new mathematical ideas to the motivating questions and issues of the domain science.

It was from this challenge facing us that the idea for the *Frontiers in Applied Dynamics* was born. Our hope is that through the editions of this series, both new and seasoned dynamicists will be able to get into the applied areas that are defining modern dynamical systems. Each chapter will expose an area of current interest and excitement, and provide a portal for learning and entering the area. Occasionally, we will combine more than one paper in a volume if we see a related audience as we have done in the first few volumes. Any given paper may contain new ideas and results. But more importantly, the papers will provide a survey of recent activity and the necessary background to understand its significance, open questions, and mathematical challenges.

Providence, RI, USA   Christopher K.R.T. Jones
Providence, RI, USA   Björn Sandstede
New York City, NY, USA   Lai-Sang Young

Preface

Natural processes can often be modeled by partial differential equations. In many applications, it is the emergence of spatially localized solutions that is of particular interest: examples are localized light pulses in photonic crystals, vortices in fluid flow, and large-scale circulation events in meteorological and climate systems. Dynamical-systems theory provides a number of techniques that can be utilized to study the existence of these coherent structures and to investigate their local and global stability properties. The way in which these techniques are used depends fundamentally on the nature of the underlying partial differential equations: the analysis of dissipative equations such as the Navier-Stokes equations differs drastically from analyses of dispersive equations that typically conserve an energy functional.

In this volume, Eugene Wayne and Michael I. Weinstein illustrate the applicability of dynamical-systems approaches in the context of dissipative and dispersive partial differential equations, respectively. Wayne reviews recent results on the global dynamics of the two-dimensional Navier-Stokes equations. This system exhibits self-similar, explicitly computable, vortex solutions. By combining classical techniques from dynamical systems theory, such as Lyapunov functions and invariant manifold theorems, one can prove that any solution of the equations for integrable initial vorticity will asymptotically approach one of these vortices - in other words, they are globally stable. However, both numerical investigations and experimental results show that in addition to the viscous time scale over which the stability of these vortices manifests itself, there are additional time scales on which important transient phenomena become evident. Wayne also surveys recent results on these metastable phenomena using analysis which originated in kinetic theory. Weinstein considers the dynamics of localized states in nonlinear Schrödinger/Gross-Pitaevskii equations play a central role in the mathematical study of nonlinear optical phenomena as well as macroscopic quantum systems, e.g. Bose-Einstein condensation. In this contribution, Weinstein reviews recent results on the bifurcation of solitary waves, their linear and nonlinear stability properties, as well as nonlinear scattering results where a conservative dissipation mechanism, radiation damping of energy to spatial infinity, plays an important role. The chapters, written independently, are combined in one volume as the Editors-In-Chief believed it would be of interest to the audience of this volume to showcase the tools of dynamical systems theory at work in explaining qualitative phenomena associated with two classes of partial differential equations with very different physical origins and mathematical properties.

Boston, MA, USA   C. Eugene Wayne
New York, NY, USA   Michael I. Weinstein

Contents

Chapter 1
Dynamical Systems and the Two-Dimensional Navier-Stokes Equations

C. Eugene Wayne

1 Introduction

The focus of this chapter is on the application of dynamical systems ideas to the study of dissipative partial differential equations with a particular focus on the two-dimensional Navier-Stokes equations. The notion of dissipativity arises in physics where it is generally thought of as a dissipation of some "energy" associated with the system and such systems are contrasted with energy conserving systems like Hamiltonian systems. In finite dimensional systems, the notion of dissipativity is relatively easy to quantify. If we have a system of ordinary differential equations (ODEs) defined on $\mathbb{R}^n$,

$$\dot{x}_j = f_j(\mathbf{x}) = f_j(x_1, \ldots, x_n) \ , \ \ j = 1, \ldots n \ , \tag{1.1}$$

then a common definition of dissipativity is that there be some bounded set, $\mathcal{B}$, which is forward invariant under the flow defined by our differential equations and such that every solution of the system of ODEs eventually enters $\mathcal{B}$, [Hal88]. Such a set is referred to as an absorbing set. If an absorbing set exists, and if ϕ^t is the flow defined by this dynamical system, then we can define an attractor for the system by

$$\mathcal{A} = \cap_{t \geq 0} \phi^t(\mathcal{B}) \ , \tag{1.2}$$

which will be compact and invariant, and will have the property that any trajectory will approach this set as $t \to \infty$.

C.E. Wayne (✉)
Department of Mathematics and Statistics, Boston University, Cummington Hall 111, Boston, MA 02215, USA
e-mail: cew@math.bu.edu

© Springer International Publishing Switzerland 2015
C.E. Wayne, M.I. Weinstein, *Dynamics of Partial Differential Equations*,
Frontiers in Applied Dynamical Systems: Reviews and Tutorials 3,
DOI 10.1007/978-3-319-19935-1_1

1

Another property often associated with dissipativity is that the determinant of the Jacobian of the vectorfield in (1.1) is negative, i.e.

$$\det\left(\{\frac{\partial f_j}{\partial x_k}\}_{j,k=1,\ldots,n}\right) < 0 . \tag{1.3}$$

If this determinant is zero, then the system conserves phase space volumes, and this is one of the properties associated with Hamiltonian systems. Condition (1.3) means that the systems "dissipates" phase space volume, but note that it is not, in itself, sufficient to insure that we have a bounded attractor for the system, as the two-dimensional example

$$\dot{x}_1 = \frac{1}{2}x_1 , \quad \dot{x}_2 = -x_2 , \tag{1.4}$$

for which almost every solution tends to infinity shows. However, if the system satisfies (1.3), and is in addition dissipative in the sense described above, then we can immediately conclude that the attractor for the system has zero n-dimensional volume. Thus, the asymptotic behavior of the system is determined by what happens on a very "small" set.

One important direction of research in the study of dissipative systems is to focus on the properties of the attractor. In special cases, the attractor may consist of a small number of simple orbits, like stationary solutions and their connecting orbits, or periodic orbits. In other, more complicated cases, the attractor may contain chaotic trajectories but it may itself live in a manifold of much lower dimension than the number of degrees of freedom of our original system. As we will see in the subsequent sections, both of these types of behavior occur in the Navier-Stokes equations, depending on whether or not an external force is present. In either case though, the long-time behavior of arbitrary solutions of the original system can be determined from a study of the possibly much smaller system obtained by restricting the original ODEs to this manifold containing the attractor. This dimensional reduction has been a powerful tool in the study of dissipative systems.

When one turns from ODEs to partial differential equations (PDEs), the discussion becomes more complicated due to the infinite dimensional nature of the problem. For one thing, the long-term behavior of the system may well depend on the norm we choose on our space of solutions. However, even after one has fixed the norm on the system the situation can be problematic due to the fact that closed and bounded sets need no longer be compact. Even if we find some bounded absorbing set $\mathcal{B}$, as above, we have no guarantee that $\cap_{t \geq 0}\phi^t(\overline{\mathcal{B}})$ will be non-empty! Thus, in addition to proving that the PDE is well-posed, one typically needs to establish some smoothing properties to show that not only is there a set $\mathcal{B}$ that eventually "absorbs" all trajectories but also that this set is precompact in the function space on which we are working.

The reduction of dimension of the problem which results from focussing one's attention on the restriction of the system to the attractor is an even more powerful tool in the context of PDEs than ODEs. As we will see in subsequent sections, many physically interesting PDEs have attractors of finite (sometimes even small) dimension. Thus, the complex behavior of solutions of the PDE can be captured by the behavior of the solutions restricted to this finite dimensional set. In one sense this simply transfers the problem from the study of the infinite dimensional behavior of solutions of the PDE to computing the possibly very complicated structure of the attractor and the latter question remains an active and open area of research for many of even the most natural physical systems, but it is at least a finite dimensional question and one that is well suited to attack with the methods of dynamical systems theory.

One may wonder how hard it is to establish that an infinite dimensional dynamical has a finite dimensional attractor. Often the smoothing and boundedness needed to prove this follows in a relatively straightforward way from the same sorts of estimates that yield existence and uniqueness of solutions. Consider, for example, the family of nonlinear heat-equations

$$u_t = u_{xx} + g(u) , 0 < x < L \tag{1.5}$$

with zero boundary conditions - i.e. $u(0, t) = u(L, t) = 0$. If one places mild growth conditions on the nonlinear term g, this is known as the Chafee-Infante equation and was one of the first PDEs to be systematically studied with the methods of dynamical systems theory. If one assumes that the initial conditions $u_0 \in H_0^1(0, L)$, then the methods of semigroup theory readily show that the orbit $u(\cdot, t)$ with this initial condition is relatively compact in this Hilbert space [Mik98], and one has an absorbing set. Hence the equation defines a dissipative dynamical system in the sense of the above definition. In this case the attractor is quite simple - for a special class of nonlinear terms, Henry showed that it consisted of the stationary solutions of the equation, and their unstable manifolds, [Hen81].

Of course, the converse of this fact is also true - there are equations like the three-dimensional Navier-Stokes equation which are expected to be dissipative on physical grounds, but the fact that there is no proof that smooth solutions exist for all time for general initial data precludes proving that they have a finite dimensional attractor.

The main example of an infinite dimensional dissipative system that we'll examine in the remainder of this review is the two-dimensional Navier-Stokes equation (2D NSE). These equations describe the evolution of the velocity of a two-dimensional fluid. While it may seem unrealistic to study two-dimensional fluid flows in a three-dimensional world there are a number of circumstances (e.g., the Earth's atmosphere) where this is a reasonable physical approximation. For further discussion of this point, see [Way11].

Physically, the NSE arises from an application of Newton's law for the fluid, namely

$$\frac{d}{dt}(\text{momentum}) = \text{applied forces} . \tag{1.6}$$

If $\mathbf{u}(\mathbf{x}, t)$ is the fluid velocity, and if we assume that the fluid is incompressible so that we can take the density to be constant, the time rate of change of the momentum is given by the convective derivative

$$\frac{d}{dt}(\text{momentum}) = \rho\frac{\partial\mathbf{u}}{\partial t} + \rho\mathbf{u}\cdot\nabla\mathbf{u}\,. \tag{1.7}$$

The second term in this expression reflects the fact that the momentum of a small region of the fluid can change not only due to changes in its velocity, but also because it is being simultaneously swept along by the background flow.

The forces are typically split into three parts:

- forces due to pressure: $f_{\text{pressure}} = -\nabla p(x, t)$, where p is the pressure in the fluid.
- viscous forces: These involve modeling internal properties of the fluid. We will take a standard model which says $f_{\text{visc}} = \alpha\Delta\mathbf{u}$, for some constant α.
- external forces, which we denote by $\mathbf{g}$.

If we insert these expressions into Newton's law, we obtain

$$\partial_t\mathbf{u} + \mathbf{u}\cdot\nabla\mathbf{u} = (\alpha/\rho)\Delta\mathbf{u} = \frac{1}{\rho}\nabla p + \frac{1}{\rho}\mathbf{g}. \tag{1.8}$$

Note that if we consider a fluid moving in d dimensions, this expression is actually a system of d partial differential equations. However, we have $d + 1$ unknown functions - the d components of the velocity, plus the pressure. To close our system we append the additional equation

$$\nabla\cdot\mathbf{u} = 0\,, \tag{1.9}$$

which reflects the assumption that the fluid is incompressible. For a further discussion of the physical origin of these equations, one can consult [DG95].

As remarked above, in three-dimensions it is not known whether or not the NSE possess smooth solutions for all times, even if one assumes that the initial velocity field is very smooth and there are no external forces acting on the system. Indeed, this is one of the famous Millennium Prize Problems. Thus, we will discuss only two-dimensional fluids, i.e. we will assume that

$$\mathbf{u} = \mathbf{u}(\mathbf{x}, t) \in \mathbb{R}^2\,, \text{ for } \mathbf{x} \in \Omega \subset \mathbb{R}^2\,. \tag{1.10}$$

In order to complete the specification of the problem, we must supplement equations (1.8)–(1.9) with appropriate boundary conditions. We'll focus on two special cases which are especially amenable to mathematical analysis, namely either

1. $\Omega = \mathbb{R}^2$, with boundary conditions imposed by assuming appropriate decay conditions on $\mathbf{u}$ at infinity, or

2. $\Omega = [-\pi, \pi] \times [-(\pi\delta), (\pi\delta)] = \mathbb{T}_\delta^2$ with periodic boundary conditions, i.e. $\mathbf{u}(x_1, x_2, t) = \mathbf{u}(x_1 + 2\pi, x_2, t) = \mathbf{u}(x_1, x_2 + 2\pi\delta, t)$. Note that δ is a parameter that measures the asymmetry of our domain - it is assumed to be $\mathcal{O}(1)$ and is not necessarily small.

Very early in the study of fluid mechanics it was pointed out by Helmholtz that it was often more useful to study the evolution of the fluid's *vorticity* than its velocity. The vorticity is the curl of the velocity field - i.e. $\overline{\omega}(\mathbf{x}, t) = \nabla \times \mathbf{u}(\mathbf{x}, t)$ and in general it is a vector, like the velocity. However, in two-dimensions an important simplification occurs:

$$\overline{\omega}(x_1, x_2, t) = (0, 0, \partial_{x_1} u_2 - \partial_{x_2} u_1) = (0, 0, \omega(x_1, x_2, t)) \tag{1.11}$$

so we see that only one component of the vorticity is non-zero and we can treat it as a scalar. If we take the curl of (1.8) we see that in two dimensions, we arrive at the scalar PDE

$$\partial_t \omega + \mathbf{u} \cdot \nabla\omega = \nu\Delta\omega + f . \tag{1.12}$$

Here the parameter $\nu = \alpha/\rho$, and $f = \frac{1}{\rho}(\partial_{x_1} g_2 - \partial_{x_2} g_1)$.

One advantage of this formulation of the problem is that the pressure term has disappeared entirely from the equation. However, the price we pay is that it appears at first that the equation is no longer well defined - the velocity $\mathbf{u}$ still appears in the equation, although we have no equation for its evolution. However, we can eliminate $\mathbf{u}$ from the equation by recalling that

- $\mathbf{u}$ is divergence free, and
- ω is the curl of $\mathbf{u}$.

This means that we can reconstruct $\mathbf{u}$ from the vorticity with the aid of Biot-Savart law, which in two-dimensions takes the form

$$\mathbf{u}(\mathbf{x}, t) = B[\omega](\mathbf{x}, t) = \frac{1}{2\pi} \int \frac{\mathbf{y}^\perp}{|\mathbf{y}|^2} \omega(\mathbf{x} - \mathbf{y}, t)d\mathbf{y} = \frac{1}{2\pi} \int \frac{(\mathbf{x} - \mathbf{y})^\perp}{|\mathbf{x} - \mathbf{y}|^2} \omega(\mathbf{y}, t)d\mathbf{y} , \tag{1.13}$$

where, if $\mathbf{y} = (y_1, y_2)$, $\mathbf{y}^\perp = (-y_2, y_1)$.

If we insert this representation into (1.12), we see that we can regard the vorticity equation:

$$\partial_t \omega = \nu\Delta\omega - B[\omega] \cdot \nabla\omega + f , \tag{1.14}$$

as a nonlinear heat equation, with a quadratic, but non-local, nonlinear term. This relationship with the heat equation, and in particular the fact that this means that the vorticity of two-dimensional flows satisfies a maximum principle will be used in subsequent sections.

In the remaining sections of this review we will discuss how dynamical systems ideas can illuminate the long-time behavior of solutions of (1.14). We begin in the next section by considering the unforced Navier-Stokes equation on the entire two-dimensional plane. We will see that in this case, if the initial vorticity distribution is in $L^1(\mathbb{R}^2)$, the attractor consists of a single point - an explicit vortex solution of (1.14). We then turn to a consideration of the problem in a periodic two-dimensional domain. In the unforced case, the attractor is simply the zero solution - i.e. the solution in which the fluid is at rest. However, numerical and physical experiments indicate that the solutions nonetheless have a variety of interesting behaviors that appear (and persist for a long time) before the solution reaches its asymptotic state. We explore these behaviors in Section 3 where we discuss metastable behavior in dissipative equations. Finally, in Section 4 we discuss what happens when the equation is subject to an external forcing. In this case, one typically has a nontrivial attractor. However, while we have far less explicit information about the attractor in this case than in the unforced situation, very general estimates of Constantin and Foias [CF85] show that the attractor remains of finite dimension, and allows us to estimate that dimension in terms of the properties of the forcing function.

2   The ω-limit set of the two-dimensional Navier-Stokes equation

In this section we focus on the long-time asymptotic behavior of solutions of the *unforced* two-dimensional vorticity equation

$$\partial_t \omega = \nu \Delta \omega - \mathbf{u} \cdot \nabla \omega \tag{1.15}$$

defined on the whole plane - i.e. $\omega = \omega(\mathbf{x}, t)$, $\mathbf{x} \in \mathbb{R}^2$. At first glance, it may seem that this equation is unlikely to yield interesting dynamics - the dissipation in the equation might be expected to dampen out any nontrivial motions. However, we will see that in the case of both bounded and unbounded domains, characteristic structures emerge in the solutions which numerical investigations indicate are also important features of two-dimensional forced flows.

Clearly, $\omega \equiv 0$ is a fixed point for this equation and from a dynamical systems point of view it is natural in this circumstance to linearize about the fixed point and ask what the linearization tells us about the behavior nearby. If we linearize the 2D vorticity equation about the zero solution, the resulting linearized equation is the 2D heat equation:

$$\partial_t \omega = \nu \Delta \omega . \tag{1.16}$$

Approaching this problem from a dynamical systems perspective a natural next step would be to construct center/stable/unstable manifolds for the nonlinear equation which correspond to the eigenspaces of the linear problem with eigenvalues having zero, negative or positive real parts, respectively.

Unfortunately we immediately face a problem if we attempt to apply those ideas in the present context. One can easily compute the spectrum of the linear operator on the RHS of the heat equation using the Fourier transform and one finds that the spectrum consists of the negative real axis, up to and including the origin. Since there is no gap in the spectrum there is no way to split the phase space into "center" or "stable" parts as required in the center manifold theorem and no obvious way of identifying the modes associated with particular decay properties.[1]

A way of circumventing this difficulty emerges if we recall the form of the fundamental solution of the heat equation:

$$G_\nu(\mathbf{x}, t) = \frac{1}{4\pi \nu t} e^{-|\mathbf{x}|^2/(4\nu t)} . \tag{1.17}$$

This suggests that it may be natural to consider (1.15) not in the variables $(\mathbf{x}, t)$, but in new variables in which $\mathbf{x}$ and t are related as $\sim \mathbf{x}/\sqrt{t}$. With this in mind, we introduce new independent and dependent variables:

$$\omega(\mathbf{x}, t) = \frac{1}{(1 + \nu t)} w(\frac{\mathbf{x}}{\sqrt{1 + \nu t}}, \log(1 + \nu t)) \tag{1.18}$$

$$\xi = \frac{\mathbf{x}}{\sqrt{1 + \nu t}} , \quad \tau = \log(1 + \nu t)$$

Remark 1. *These types of variables are often used in studying parabolic PDE where they are sometimes referred to as scaling variables.*

If we now rewrite (1.16) in terms of these new variables we obtain the new PDE

$$\partial_\tau w = \mathbb{L}w , \quad w = w(\xi, t) , \quad \xi \in \mathbb{R}^2 \tag{1.19}$$

$$\mathbb{L}w = \Delta_\xi w + \frac{1}{2}\xi \cdot \nabla_\xi w + w = \Delta_\xi w + \frac{1}{2}\nabla_\xi \cdot (\xi w)$$

At first sight, it may not be apparent why (1.19) is an improvement over (1.16) as we have made the equation more, rather than less, complicated. However, as we show below, in contrast to the Laplace operator which appears on the right-hand side of the heat equation, the operator $\mathbb{L}$ has a gap in its spectrum between the part of the spectrum with zero real part and the remainder and this will allow us to apply the center-manifold theorem to understand the asymptotic behavior of solutions near the fixed point at the origin.

To understand the spectrum of $\mathbb{L}$, consider the eigenvalue problem

$$\mathbb{L}\psi_\lambda = \lambda\psi_\lambda . \tag{1.20}$$

---

[1]The situation is very different if one considers the equation on a bounded domain with periodic boundary conditions - see the discussion of the work of Foias and Saut [FS84a] in the following section.

If we separate variables in this PDE, we get a pair of ordinary differential equations of the form

$$\frac{d}{d\xi_1^2}\phi_\lambda + \frac{1}{2}\frac{d}{d\xi_1}(\xi_1\phi_\lambda) = \lambda\phi_\lambda \, , \tag{1.21}$$

with a similar equation for the ξ_2 part of the solution. Taking the Fourier transform of this equation yields

$$-k^2\hat{\phi}_\lambda - \frac{1}{2}k\frac{d}{dk}\hat{\phi}_\lambda = \lambda\hat{\phi}_\lambda \, . \tag{1.22}$$

This first order equation can be solved with the aid of integrating factors and one finds that for any $\lambda \in \mathbb{C}$ one has a solution:

$$\hat{\phi}_\lambda(k) = \frac{C^+}{|k|^{2\lambda}}e^{-k^2}H(k) + \frac{C^-}{|k|^{2\lambda}}e^{-k^2}H(-k) \tag{1.23}$$

where $H(k)$ is the Heaviside function and the fact that we have two constants of integration for a first order differential equation reflects the singular point at $k = 0$.

At first sight, this seems as if every point λ is in the spectrum of $\mathbb{L}$. However, recall that the spectrum of an operator depends on the space on which it acts. In particular, if $\Re(\lambda) > 0$, the functions in (1.23) "blow-up" at $k = 0$ and hence won't be in any "well behaved" function space. In order to say exactly what the spectrum of $\mathbb{L}$ is, we must decide what function space it acts on - like many operators, its spectrum will change, according to the domain chosen. It has long been known that the time-decay of solutions of parabolic PDEs is linked to the spatial decay rate of their solutions. With this in mind, we define a family of weighted Sobolev spaces:

$$L^2(m) = \{f \in L^2(\mathbb{R}^2) \mid \|f\|_m < \infty\}$$

$$\|f\|_m = \left(\int_{\mathbb{R}^2}(1 + |\xi|^2)^m|f(\xi)|^2d\xi\right)^{1/2} \tag{1.24}$$

$$H^s(m) = \{f \in L^2(m) \mid \partial^\alpha f \in L^2(m) \text{ for all } \alpha = (\alpha_1,\dots,\alpha_d) \text{ with } |\alpha| \le s\}$$

One reason that these spaces are so convenient for our purposes is the fact that Fourier transformation turns differentiation into multiplication and vice versa. For these spaces, that makes it easy to check that for any non-negative integers s and m, Fourier transformation is an isomorphism between $H^s(m)$ and $H^m(s)$, i.e. a function f is in $H^s(m)$, if and only if its Fourier transform $\hat{f} \in H^m(s)$.

Applying this observation to the expression for $\hat{\phi}_\lambda$ in (1.23), we see that due to the singularity at $k = 0$ and the rapid decay as $|k| \to \infty$, $\hat{\phi}_\lambda \in H^m(s)$ if the first m derivatives of $\hat{\phi}_\lambda$ are square integrable in some neighborhood of the origin.

Note that there are some "special" values of λ. If $\lambda = 0, -1/2, -1, -3/2, \dots$, we can choose $A^\pm$ so that $\hat{\phi}_{-n/2}(k) = Ak^n\exp(-k^2)$. These are entire, rapidly decaying functions and thus are elements of $H^m(s)$ for any values of m and s, so that the non-negative half integers are eigenvalues of $\mathbb{L}$ for any s and m, with eigenfunctions given by the inverse Fourier transform of these expressions. In particular, we see that $\lambda = 0$ is always an eigenvalue and its eigenfunction is the Gaussian $\phi_0(\xi) = C_0 \exp(-\xi^2/4)$. For non-half integral values of λ, we cannot choose the constants $A^\pm$ to make the eigenfunctions smooth, and thus, at least for some values of s, they will not be in the spaces $H^m(s)$. In fact, one can easily verify that in one-dimension, for no value of $\lambda \subset \mathbb{C}$ with $\Re(\lambda) \geq 0$ will $\hat{\phi}_\lambda$ be in $H^1(s)$, because the derivative will have a non-square-integrable singularity at the origin. In two-dimensions, one can tolerate slightly worse singularities, but one still finds that no ϕ_λ with $\Re(\lambda) \geq 0$ will lie in $H^2(s)$. A careful examination of this argument allows one to compute exactly what the spectrum of $\mathbb{L}$ is and one finds

Theorem 1. *([GW02], Theorem A.1)Fix $m > 1$ and let $\mathbb{L}$ be the operator in* (1.19) *acting on its maximal domain in $L^2(m)$. Then*

$$\sigma(\mathbb{L}) = \left\{ \lambda \in \mathbb{C} \mid \mathrm{Re}(\lambda) < \frac{1}{2} - \frac{m}{2} \right\} \cup \left\{ -\frac{n}{2} \mid n = 0, 1, 2, \dots \right\} .$$

The key fact here, from the point of view of dynamical systems theory, is that we have now created a spectral gap. So long as we choose the decay rate parameter $m \geq 2$ in our function space, the eigenvalue $\lambda = 0$ is an isolated point in the spectrum of $\mathbb{L}$, with all the rest of the spectrum lying strictly in the left half plane. Thus, at least intuitively, we can now hope to define the center subspace of the linear problem to be the span of the eigenfunction(s) of the zero eigenvalue, the stable subspace to be the spectral subspace corresponding to the remainder of the spectrum, and attempt to construct a center-manifold for the full nonlinear equation that it is tangent to the zero-eigenspace at the origin.

We first note that the zero-eigenspace is very simple in this case - it is one-dimensional and consists just of the span of the Gaussian function

$$\phi_0(\xi) = \frac{1}{4\pi} e^{-\xi^2/4} . \tag{1.25}$$

For future reference we note that the projection onto this eigenspace is given by the zero eigenfunction to the adjoint operator to $\mathbb{L}$, namely

$$\mathbb{L}^\dagger v = \nu \Delta v - \frac{1}{2} \xi \cdot \nabla v . \tag{1.26}$$

The constant function is clearly a zero eigenfunction for $\mathbb{L}^\dagger$ and consequently, the projection of $f \in H^2(m)$ onto the zero eigenspace of $\mathbb{L}$ is given just by $P_0 f = (\int_{\mathbb{R}^2} f(\xi) d\xi) \phi_0$.

While the spectral gap in the spectrum of the linear part of the equation makes it reasonable to expect that there will be a center manifold for the semi-flow defined by the 2D NSE, there are technical difficulties associated with constructing invariant manifolds for PDE that still must be overcome. In contrast to the case of ordinary differential equations where it is more or less clear what the optimal assumptions on the vectorfield should be in order to obtain an invariant manifold theorem, the situation is far less clear in the case of PDE. For instance, depending on the circumstances, it may be preferable to assume that the semi-group associated with the linear part of the equation may be more or less smoothing. Depending on the choices made in this case, one may need to either make stronger hypotheses about the nonlinear part of the equation or draw weaker conclusions about the manifold one constructs. There is far less of a "one size fits all" invariant manifold theorem in infinite dimensional systems - instead one typically tailors the hypotheses of the theorem to the circumstances of interest. For some of the choices that have been made, see [BJ89, Mie91] or [VI92]. One general principle that seems to emerge from these different contexts is that it is often easier to work with the semi-flow defined by the PDE, rather than the equation itself, if this semi-flow exists. This is because it already incorporates any smoothing associated with the linear evolution, and we don't need to worry about precisely what smoothing assumptions to make on the flow. (For instance, one common assumption is that the linear part of the equation defines an analytic semi-group, but while such a hypothesis would apply to the vorticity equation (1.12), it would no longer hold once we rewrite the equation in terms of the scaling variables (1.18).)

One infinite dimensional version of the center-manifold theorem that does apply to our problem is the version due to Chen, Hale, and Tan (CHT) [CHT97]. (CHT) assume that the PDE defines a semi-flow ϕ^t on some Banach space, X. They then make four natural hypotheses about this semi-flow. We refer the reader to the original paper for the details on these assumptions, but roughly they are as follows:

(H1) $\phi^t(u)$ is Lipshitz continuous in both t and u with uniformly bounded Lipshitz constant for t in some interval.

(H2) For some fixed, positive τ, ϕ^τ can be split into a bounded linear operator and a globally Lipshitz operator.

(H3) The Banach space can be split into a direct sum of two pieces (a "center" subspace, X_1 and a "stable" subspace X_2) and when the linear part of ϕ^τ acts on the stable subspace it produces decay at a faster rate than the growth produced when the inverse of the linear part of ϕ^τ acts on the center subspace. (This is a reflection and consequence of a spectral gap in the spectrum of the linear part of the PDE.)

(H4) The ratio of the Lipshitz constant of the nonlinear term divided by the spectral gap must be small.

Assuming that these hypotheses are satisfied, (CHT) then prove the following:

• There exists a Lipshitz function $g : X_1 \rightarrow X_2$ whose graph is invariant with respect to ϕ^τ. This is the center manifold for our system.

• Any orbit of ϕ^τ will approach an orbit on the center manifold as time goes to infinity.

We will apply the (CHT) theorem to the two-dimensional vorticity equation, (1.15), rewritten in terms of the scaling variables (1.18). In addition to redefining the dependent vorticity variable ω as in (1.18) we also rescale the velocity variable as

$$\mathbf{u}(\mathbf{x}, t) = \frac{1}{\sqrt{1 + vt}} \mathbf{v}(\frac{\mathbf{x}}{\sqrt{1 + vt}}, \log(1 + vt)) . \qquad (1.27)$$

Remark 2. *With this definition one can check that the rescaled vorticity w and rescaled velocity* $\mathbf{v}$ *are still related to one another through the Biot-Savart law.*

If we rewrite (1.15) in terms of these new variables, we find

$$\partial_\tau w = \mathbb{L}w - \frac{1}{v}\mathbf{v} \cdot \nabla_\xi w . \qquad (1.28)$$

If we take initial condition $w|_{\tau=0} = w_0$, we can define the semi-group associated with (1.28) with the aid of the variation of constants formula as

$$w(\tau) = e^{\tau \mathbb{L}} w_0 - \frac{1}{v} \int_0^\tau e^{(\tau-s)\mathbb{L}}(\mathbf{v}(s) \cdot \nabla_\xi w(s)) dx . \qquad (1.29)$$

The two terms in (1.29) define the splitting into linear and nonlinear pieces in hypothesis ($H2$), and the decay hypotheses of ($H3$) are intuitively satisfied due to the gap in the spectrum of $\mathbb{L}$. One can check them rigorously just by noting that the semigroup $e^{t\mathbb{L}}$ is just the heat semigroup written in terms of the scaling variables. Expressing the heat semigroup in terms of these variables then leads to the expected estimates, [GW02].

The remaining hypotheses ($H1$) and ($H4$) as well as the second part of ($H2$) concern the smoothness of the nonlinear term. In order to treat this term, one needs estimates that allow us to transfer information about the vorticity field to information about the velocity field. The types of estimates we will need are analogous to the Hardy-Little-Sobolev (HLS) inequality, [LL97]. Note that from (1.13), we have

$$|v_j(\mathbf{x})| \le \frac{1}{2\pi} \int_{\mathbb{R}^2} \frac{1}{|\mathbf{x} - \mathbf{y}|} |w(\mathbf{x} - \mathbf{y})| d\mathbf{y} . \qquad (1.30)$$

Thus, if $w \in L^p(\mathbb{R}^2)$, the HLS inequality immediately implies that

$$\|u\|_{L^q} \le C\|\omega\|_{L^p} , \qquad (1.31)$$

for $q = 2p/(2-p)$. We need to refine these inequalities to account for the weights in our function spaces (i.e., to account for the fact that we work not in L^2, but in $L^2(m)$), but very similar estimates hold in that case. An extensive set of such inequalities is proved in ([GW02], Appendix B).

These inequalities allow one to establish the Lipshitz bounds used in ($H1$) and ($H4$). However, they would not immediately yield the *global* Lipshitz estimate required in the second part of (H2). This is obtained as in other constructions of the center-manifold theorem by "cutting off" the nonlinear term outside of some ball centered at the origin. More precisely we multiply the nonlinear term by $\chi(\|w\|_{L^2(m)})$, where χ is a smooth function equal to one on a neighborhood of zero and vanishing outside of some slightly larger neighborhood. This results in an equation whose nonlinear term now has a global Lipshitz bound and whose solutions agree with those of our original equation whenever $\|w\|_{L^2(m)}$ is sufficiently small. Again, the details are provided in [GW02].

The estimates above allow us to apply the (CHT) theorem to construct a one-dimensional center manifold for the two-dimensional vorticity equation, written in terms of the scaling variables as in (1.28). Furthermore, the (CHT) theorem also guarantees that all solutions of the equation which remain in some neighborhood of the origin will approach an orbit on the center manifold. (We note that this restriction to solutions near the origin is not contained in the statement of the (CHT) theorem above. It results from the fact that we had to cut off the nonlinear term in (1.28) in order to verify the hypotheses of the theorem, and hence the conclusions of the theorem only apply to solutions that remain in the region where the cutoff function is one.) Thus, all the long-time asymptotics of such solutions will be determined by the orbits on the center manifold, so our next step is to compute the dynamics of (1.28) when restricted to this manifold. In the vorticity equation the center subspace X_1 is just the span of the eigenfunction of the zero eigenvalue which we showed above is the Gaussian function ϕ_0 - i.e. any function $w_c \in X_1$ has the form $\alpha\phi_0$. Thus, any point in the center manifold can be written as $w = \alpha\phi_0 + g(\alpha)$, where g is the function whose graph defines the center manifold, and the restriction of (1.28) to the center manifold can be written as

$$\dot{\alpha}\phi_0 = \mathbb{P}_0(N(\alpha\phi_0 + g(\alpha))) \,. \tag{1.32}$$

Here, N is shorthand notation for the nonlinear term in (1.28), $\mathbb{P}_0$ is the projection onto the center subspace, and there is no linear term in the equation because the eigenvalue corresponding to ϕ_0 which would give rise to the linear term in the equation is zero. Recall that as we showed just after (1.26), the projection of a function f onto the center subspace is given by multiplying by ϕ_0 by the integral of f over $\mathbb{R}^2$. Thus, we need to integrate the nonlinear term in (1.28) over $\mathbb{R}^2$ - i.e.

$$-\frac{1}{\nu}\int_{\mathbb{R}^2}(\mathbf{v}\cdot\nabla w)d\xi \tag{1.33}$$

But if we recall that $\mathbf{v}$ is incompressible, i.e. has zero divergence, we can rewrite the integrand in this expression as $\nabla \cdot (\mathbf{v}w)$, and then the integral vanishes due to the divergence theorem! Thus, the restriction of (1.28) to the center manifold is simply

$$\dot{\alpha} = 0 \,. \tag{1.34}$$

This implies the surprising fact that the center manifold consists entirely of fixed points! Even more surprising, we can write them down explicitly - they are just multiples of the Gaussian. That is, if we take

$$w(\xi, \tau) = \mathcal{O}^A(\xi, \tau) = \frac{A}{4\pi} e^{-|\xi|^2/4} \tag{1.35}$$

this is a fixed point of (1.28) for all values of A. The reason is that if one computes the velocity field corresponding to $\mathcal{O}^A$ from the Biot-Savart law it is

$$\mathbf{v}^A(\xi, \tau) = \frac{A}{2\pi} \frac{\xi^\perp}{|\xi|^2} \left(1 - e^{-|\xi|^2/4}\right). \tag{1.36}$$

Note that this velocity field is a purely tangential vector field, while $\mathcal{O}^A$ is purely radial and as a consequence, $\mathbf{v}^A \cdot \nabla \mathcal{O}^A \equiv 0$. Recalling that $\mathbb{L}\phi_0 = 0$, since ϕ_0 is the eigenfunction with eigenvalue zero, we see that the RHS of (1.28) vanishes when evaluated at $\mathcal{O}^A$ and as a consequence these are all fixed points of the equation.

Remark 3. *These solutions of the two-dimensional Navier-Stokes equation have been known for a long time and are called Lamb-Oseen vortices. If expressed in terms of the original variables instead of the scaling variables, they are self-similar solutions of the equation, rather than fixed points.*

Remark 4. *We can use the (CHT) theorem to construct other invariant manifolds, besides just the center manifold. Recall the computation of the spectrum in Theorem 1. If we choose the decay rate m of our function space $m \geq 3$, we see that there are now two isolated eigenvalues in addition to the essential spectrum - namely $\lambda_0 = 0$ and $\lambda_1 = -1/2$. If one examines the eigenfunctions computed above, one finds that $-1/2$ is a double eigenvalue and its eigenfunctions are Hermite functions, $C_1 \xi_1 e^{-|\xi|^2/4}$ and $C_1 \xi_2 e^{-|\xi|^2/4}$, where the constant C_1 is chosen to normalize the eigenfunctions. One can then reapply the (CHT) theorem, this time taking as the "center" subspace X_1, the three-dimensional space spanned by the Gaussian and these two Hermite functions. This gives rise to a three-dimensional invariant manifold, and just as above one can compute explicitly the system of ordinary differential equations that results when one restricts (1.28) to this manifold. This computation uses the fact that the eigenfunctions of the adjoint operator $\mathbb{L}^\dagger$ with eigenvalues $-1/2$ are just the coordinate functions ξ_1 and ξ_2 and hence the projection onto these directions corresponds just to taking first moments of the solution. By choosing larger and larger values of m, one can expose more and more isolated eigenvalues, and one finds that their eigenfunctions are also given by Hermite functions of higher and higher order, and the projections onto these eigendirections are given by combinations of higher order moments of the solution.*

Remark 5. *There has been a fair amount of work using more traditional PDE techniques to identify special families of solutions of the Navier-Stokes equations which have specific temporal decay properties. For instance, Mayakawa and*

Schonbek [MS01] gave necessary and sufficient conditions for solutions to satisfy specific temporal decay rates in terms of integrals of various moments of the solutions. From the dynamical systems point of view, we see that solutions decaying with a specific rate can be identified as lying in a particular invariant manifold. For example, any solution approaching a non-zero Oseen vortex in the center manifold will decay in time with the same rate as the Oseen vortex, i.e. $\sim t^{-1/2}$ in the L^∞ norm in the original, unscaled variables. The only way a solution in a neighborhood of the origin can avoid approaching one of the Oseen vortices is if it lies in the invariant manifold of solutions asymptotic to the origin. (These manifolds are sometimes called Fenichel fibers, and they consist of all solutions sharing the same long-time asymptotics.) As we noted in the preceding remark, the projections onto the various eigenspaces are expressed as moments of the solution. Gallay and I showed that the condition that a solution lay on the Fenichel fiber through the origin was exactly the same condition on the moments that had earlier been found by analytic means in [MS01].

Remark 6. *In Remark 4 we observed that the Hermite functions are eigenfunctions corresponding to the isolated eigenvalues of $\mathbb{L}$ and these can be used to construct invariant manifolds that govern the long-time asymptotics of solutions. They can also be used as the basis of a numerical method which expresses the solution of (1.12) as a sum of finitely many vortex blobs and then replaces the PDE (1.12) with a system of ordinary differential equations that track how the centers and the moments of each of these blobs evolve, [NSUW09]. The coefficients in these ordinary differential equations are expressed in terms of integrals over Hermite functions and thanks to the integration formulas for products of Hermite functions, one can derive compact combinatorial formulas for these coefficients, allowing efficient numerical implementation of these equations [UEWB12].*

The invariant manifold approach above has allowed us to identify the family of Oseen vortices as the only candidates for the long time asymptotic behavior of *small* solutions of the 2D NSE. However, there is still the possibility that if one chooses large initial data, some other type of behavior might emerge. In order to investigate that possibility we turn to a more global tool from dynamical systems, namely Lyapunov functions. Recall that, roughly speaking, Lyapunov functions are functions defined on the phase space of a dynamical system which are monotonic non-increasing along orbits of the system. Because of this monotonicity, we see that if an orbit approaches some long-time limit, the Lyapunov function, evaluated along that orbit, must also approach a limit and hence the orbit must approach a region of the phase space in which the Lyapunov function is constant. This last observation is the heart of the LaSalle Invariance Principle, and it is extremely useful in pinning down the possible locations of the ω-limit set of a dynamical system.

We now make these observations more precise.

Definition 1. *If X is a Banach space, a Lyapunov function for the semi-flow ϕ^t is a continuous, real-valued function Ψ, such that*

$$\limsup_{t \to 0^+} \frac{\Psi(\phi^t(u_0)) - \Psi(u_0)}{t} \leq 0 \text{ for all } u_0 \in X . \tag{1.37}$$

The LaSalle Invariance Principle can then be stated as:

Proposition 1. *Let Ψ be a Lyapunov function for the semi-flow ϕ^t. Define $\mathbb{E} = \{u \in X \mid \frac{d}{dt}\Psi \cdot \phi^t(u)|_{t=0} = 0\}$. If the forward orbit of u_0 is contained in a compact subset of X, then the ω-limit set of u_0 lies in $\mathbb{E}$.*

Proof. The proof makes precise the idea sketched above. The compactness of the forward orbit of u_0, plus the continuity of Ψ means that $\Psi(\phi^t(u_0))$ is bounded below as a function of t. The monotonicity of Ψ along orbits then implies that there exists Ψ^∞ such that $\lim_{t \to \infty} \Psi(\phi^t(u_0)) = \Psi^\infty$. If we choose any point w in the ω-limit set, then there exists a sequence of times t_n tending toward infinity such that $\phi^{t_n}(u_0) \to w$, and this, combined again with the continuity of Ψ, means that $\Psi(w) = \Psi^\infty$. Since the ω-limit set is invariant under ϕ^t, and since w was an arbitrary point in the ω-limit set, we find that $\Psi(\phi^t(w)) = \Psi^\infty$ for all t and thus $w \in \mathbb{E}$.   □

We now apply to the method of Lyapunov functions to the 2D NSE. We'll continue to work with the vorticity form of the equation, and since we're particularly interested in the Oseen vortices as possible ω-limit sets for solutions of this equation, we'll also continue to use the rescaled form of the equation (1.28).

The first thing we address is whether or not the forward orbit of a general initial condition w_0 is compact, since this plays an important role, both in the existence of the ω-limit set and in the LaSalle Invariance Principle. In fact, the first question to address is whether or not solutions even exist for general initial data. This turns out to be relatively easy to establish if we work in the weighted L^2 spaces we introduced earlier due to the relationship of (1.12) to the heat equation and the well-understood smoothing and decay properties of the heat kernel. In fact, one can establish decay in much larger spaces of initial data. Work by Giga, Miyakawa and Osada, Ben-Artzi, Gallagher and Gallay, and others over the past twenty years or so has proven that the equation is globally well-posed for an initial vorticity distribution in $L^1(\mathbb{R}^2)$, or even if one takes measures as initial data, [GMO88, BA94, GG05]. We won't discuss the proofs of these results because they don't have a particularly dynamical systems "flavor" which is the focus of this review, but instead refer the reader to the original articles for details.

Suppose, given the well-posedness results of the previous paragraph that we consider an arbitrary initial vorticity distribution $w_0 \in L^1(\mathbb{R}^2)$. The forward orbit of this point exists, and we would like to know if it has an ω-limit. This will follow if the orbit is relatively compact in L^1, and by the Rellich compactness criterion, this will in turn follow if we can demonstrate that the solutions have some smoothness and decay at infinity. In the case of (1.28):

- the smoothness of the solution comes from the smoothing properties of the heat kernel, which are preserved by the nonlinear term in the equation, and

- the decay at infinity comes from estimates on solutions of the vorticity equation due to Carlen and Loss [CL95].

The details of this argument are presented in [GW05], and establish that the ω-limit set exists for any solution of (1.28) with initial vorticity in L^1.

We now compute what the ω-limit set actually is with the aid of two Lyapunov functions.

(A) The first Lyapunov function is motivated by relative entropy functions of kinetic theory. In kinetic theory, one often seeks to prove that the probability distribution for the velocities of a gas of particles converge to the Maxwellian distribution, i.e. a Gaussian distribution of the velocities. In our case, if the Oseen vortex is indeed the ω-limit set, then we also are looking for convergence to a Gaussian distribution - in this case of vorticity, rather than velocity.

(B) A significant problem with the analogy between kinetic theory and the vorticity equation is that while it is very natural to assume that the solutions of kinetic equations are non-negative (since they represent probability distributions) it is quite unnatural to assume that the vorticity is always of one sign. Since the relative entropy functional is only defined for functions that are everywhere positive (or negative), our second Lyapunov functional will ensure that even for solutions of (1.28) which change sign, the ω-limit set will still lie in the space of solution that are everywhere positive or everywhere negative.

We first focus on the relative entropy function from kinetic theory. The entropy functional, which originated in the study of statistical physics, is given by $\int w(x) \ln(w(x))dx$. The relative entropy function modifies this to look at the entropy relative to some fixed state - in our case the Oseen vortex. Thus, we define

$$H(w(\tau)) = \int_{\mathbb{R}^2} w(\xi, \tau) \ln\left(\frac{w(\xi, \tau)}{\phi_0(\xi)}\right) d\xi , \qquad (1.38)$$

where ϕ_0 is the Gaussian function defined in (1.25). A straightforward computation shows that H is defined, continuous, and bounded below for any function $w \in L^2(m)$ which is everywhere positive, if $m > 3$. A similar definition can be constructed for everywhere negative functions, but it is not obvious how this functional can be modified to accommodate solutions that change sign. Differentiating $H(w(\tau))$ with respect to τ, we find

$$\frac{d}{d\tau}H(w(\tau)) = \int_{\mathbb{R}^2} w_\tau \left(1 + \ln\left(\frac{w(\tau)}{\phi_0}\right)\right) d\xi . \qquad (1.39)$$

If one now inserts the expression for w_τ from the RHS of (1.28) and integrates by parts (repeatedly!) one finds that

$$\frac{d}{d\tau}H(w(\tau)) = -\int_{\mathbb{R}^2} w \left|\nabla\left(\ln\frac{w}{\phi_0}\right)\right| d\xi . \qquad (1.40)$$

Since $w(\xi, \tau) > 0$, this calculation implies that H is strictly decreasing unless the $\left| \nabla \left(\ln \frac{w}{\phi_0} \right) \right| = 0$, that is, unless $w = A\phi_0$ for some constant A. But then, by the LaSalle Invariance Principle, the ω-limit set of the orbit $w(\tau)$ must lie in the set of functions proportional to ϕ_0 - i.e. the ω-limit set must be one of the Oseen vortices. Thus, we have established that for solutions of (1.28) which do not change sign, the ω-limit set, must be one of the Oseen vortices, regardless of the size of the initial data, and we now turn to a consideration of what to do when the solution changes sign.

Remark 7. *In the calculation above, we used that if $\left| \nabla \left(\ln \frac{w}{\phi_0} \right) \right| = 0$, then $w = A\phi_0$. In principle, the constant A could depend on τ. This cannot occur in our context because of the fact that $\int_{\mathbb{R}^2} w(\xi, \tau)d\xi$ is constant. Hence the total "mass" of the solution is conserved and A cannot change with time. Note that we do assume in this calculation that the initial conditions are chosen so that $\int_{\mathbb{R}^2} w_0(\xi)d\xi \neq 0$.*

In order to treat solutions of (1.28) that change sign we exploit the similarity of the vorticity equation to the heat equation and in particular, we use the fact that its solutions satisfy a maximum principle. Given an initial condition ω_0 for (1.12) (or w_0 for (1.28)), split it into its positive and negative pieces - i.e. define

$$\omega_0^+(\mathbf{x}) = \max(\omega_0(\mathbf{x}), 0)$$
$$\omega_0^-(\mathbf{x}) = -\min(\omega_0(\mathbf{x}), 0) .$$

Then define the evolution of the positive and negative parts of the data by

$$\partial_t^{\pm} \omega = \nu \Delta \omega^{\pm} - \mathbf{u} \cdot \nabla \omega^{\pm} . \tag{1.41}$$

Then if $\omega(\xi, t)$ is the solution of (1.12), with initial condition ω_0, we have

- $\omega(\mathbf{x}, t) = \omega^+(\mathbf{x}, t) - \omega^-(\mathbf{x}, t)$, and

- Both ω^+ and ω^- satisfy a maximum principle. In particular, since $\omega_0^{\pm}(\mathbf{x}) \geq 0$, we have $\omega^{\pm}(\mathbf{x}, t) > 0$ for all $\mathbf{x}$ and $t > 0$.

With these observations, it is easy to show that the L^1 norm of ω is a Lyapunov functional (for the details of this calculation, see [GW05].) Namely, we have

Lemma 1. *Define $\Phi(\omega(t)) = \int_{\mathbb{R}^2} |\omega(\mathbf{x}, t)| \, d\mathbf{x}$. Then $\Phi(\omega(t))$ is non-increasing in time, and is strictly decreasing unless $\omega(\mathbf{x}, t)$ is everywhere positive or everywhere negative.*

Putting together our two Lyapunov functionals we can now show that for any solution of (1.28), the ω-limit set must be an Oseen vortex. Let Ω be the ω-limit set of a solution of (1.28) with initial condition $w_0 \in L^1(\mathbb{R}^2)$. Assume that $\int_{\mathbb{R}^2} w_0(\xi)d\xi \neq 0$. Applying the LaSalle Invariance Principle to the Lyapunov functional Φ, we see that any point Ω must lie in the set of functions which are everywhere positive or everywhere negative. But then, pick a point $\bar{\omega} \in \Omega$ and apply the LaSalle Invariance Principle again, this time with the relative entropy functional. From this we conclude that ω-limit set must be of the form $A\phi_0$, with $A = \int_{\mathbb{R}^2} w_0(\xi)d\xi$, and so the ω-limit set of every solution in L^1 is just an Oseen vortex.

Remark 8. *Note that there's one additional step that we have swept under the rug here. We only know that the relative entropy functional is continuous and bounded on the weighted Hilbert spaces, $L^2(m)$, not on all of L^1, so we can't directly apply the above argument to solutions in L^1. However, using the decay estimates of Carlen and Loss mentioned above, one can prove that the ω-limit set of any L^1 solution must lie in the spaces $L^2(m)$ for any $m > 1$, and then one can repeat the above argument.*

To conclude this section note that we have now shown that any solution of the two-dimensional NSE (with integrable, nonzero total vorticity) will eventually approach an Oseen vortex. If we start with small initial data, the invariant manifold theorem gives us very precise information about the asymptotic rate of approach of the solution to the Oseen vortex, but for general initial data, it may take a very long time for the solution to approach this limiting state. In the next section of this review, we examine some possible behaviors that may occur on intermediate time scales, before the solution finally converges to its asymptotic state.

3   Metastable states, pseudo-spectrum and intermediate time scales

In this section we look at another application of dynamical systems ideas to the two-dimensional NSE, namely the emergence of metastable states in the system. That part of this section which is original work is all joint work with Margaret Beck, and the details of the proofs appear in [BW11a]. In contrast to the previous section we now consider the equations on a rectangular domain with periodic boundary conditions. We are specifically interested in this section in the appearance of structures *before* the long-time asymptotic state appears, and most of the numerical studies of these phenomena have been done on such periodic domains. As in the previous section, it is convenient to study the evolution of the vorticity

$$\partial_t \omega = \nu \Delta \omega - \mathbf{u} \cdot \nabla \omega \,, \tag{1.42}$$

but this time we require that

$$\omega(x_1, x_2, t) = \omega(x_1 + 2\pi, x_2, t) = \omega(x_1, x_2 + 2\pi\delta, t) \,, \tag{1.43}$$

where $\delta \sim \mathcal{O}(1)$ is the asymmetry parameter of the domain (and will equal one for a square domain.) As in the previous section we can recover the velocity field in (1.42) from the vorticity via the Biot-Savart law, which in this case is most conveniently expressed in terms of the Fourier coefficients of the solution:

$$\hat{\omega}(k, \ell) = \frac{1}{4\pi^2 \delta} \int_{\mathbb{T}_\delta^2} \omega(x_1, x_2) e^{-i(kx_1 + \ell x_2/\delta)} dx_1 dx_2 , \qquad (1.44)$$

with analogous definitions of $\hat{\mathbf{u}}_{1,2}(k, \ell)$. (To save space, we suppress the time dependence of the functions when it will not cause confusion.) The Biot-Savart law then takes the form

$$\hat{\mathbf{u}}(k, \ell) = i \frac{(-\ell/\delta, k)}{k^2 + (\ell/\delta)^2} \hat{\omega}(k, \ell) . \qquad (1.45)$$

Remark 9. *We leave it as an easy exercise to show that because of the periodic boundary conditions, $\hat{\omega}(0, 0) = 0$, so that (1.45) is well defined. One could choose $\hat{\mathbf{u}}(0, 0)$ to be an arbitrary constant, but we will set it equal to zero. Given (1.45), one can derive estimates on the norm of the velocity in terms of those of the vorticity, analogous to those in Section 2.*

Note that from (1.45), we see immediately that if $\omega \in L^2(\mathbb{T}_\delta^2)$, then so are both components of the velocity field. If we apply the energy inequality derived below in (1.72), and take advantage of the fact that the external forcing is zero here, we see that all solutions will tend asymptotically to zero. We note here an important distinction between the Navier-Stokes equation on the torus and in the plane. In both cases, the smoothing of the evolution implies that if the initial vorticity, $\omega_0 \in L^1$, then $\omega(t) \in L^2$ for any $t > 0$. However, in the present case, as noted just above, this implies that the system has finite energy and hence will decay to zero as $t \to \infty$, rather than to an Oseen vortex, as in the previous section. One can once again use invariant manifolds to characterize the way solutions approach their asymptotic state and Foias and Saut constructed these manifolds and explored their uses in [FS84a], [FS84b]. Because the domain of the fluid is bounded, the linearization of the equation about the zero solution has discrete spectrum and it is not necessary to introduce scaling variables as we did in the case of an unbounded domain.

The Foias and Saut manifolds control the solution as it approaches its asymptotic state, and from the energy inequality, we see that the rate of convergence toward the asymptotic states occurs on the viscous time scale $t \sim \mathcal{O}(1/\nu)$. In the weakly viscous regime in which turbulent fluids are typically studied this timescale is enormously long. However, in numerical experiments on two-dimensional turbulent flows, one sees that on time scales much shorter than the viscous time scale characteristic structures emerge which then come to dominate the flow for very long periods of time, until one finally reaches the asymptotic state. The goal in the present section is to describe some recent work which proposes an explanation of the emergence of these intermediate time scales based on dynamical systems theory.

In order to gain some insight into what the metastable states and their associated time scales are in the NSE, we first review some of the numerical results on this system. One of the starting points of Beck's and my work were the investigations of Yin, Mongomery, and Clercx [YMC03]. While their numerical experiments are consistent with an eventual convergence of solutions to zero, much more striking is that characteristic structures like vortex dipole pairs or "bar states" (shear flows, in which the vorticity contours are constant in one direction) emerge quickly from an initially very disordered state, and then dominate the subsequent evolution of the flow for very long times.

In the numerics of [YMC03], the most common metastable states that are observed in the system are the dipole states - only with rather carefully prepared initial conditions are the bar states observed. However, if instead of considering the equation on a square domain as in [YMC03], one considers the equation on a rectangular domain, the numerical experiments of Bouchet and Simonnet [BS09] indicate that the bar states can become the dominant metastable states. Furthermore, these states are sufficiently stable that they continue to dominate the evolution even if the equation is subjected to a random force. While the random force may cause an apparently random switching between the bar and dipole states, for the great majority of the time, the system is in one or the other of these two states.

The goal in the remainder of this section is to propose an explanation for the rapid appearance and long persistence of these families of solutions of the two-dimensional NSE. In [BW11b], Beck and I proposed a dynamical systems explanation for similar families of metastable states in Burgers equation. These states, and their importance for the dynamics of the system, were first systematically investigated by Kim and Tzavaras in [KT01], where they are called "diffusive N-waves." Our explanation of the metastable behavior in Burgers equation began by showing that (in scaling variables, similar to those used in the previous section) there was a one-dimensional invariant manifold in the infinite dimensional phase space of the equation which is completely filled with fixed points and these fixed points represent the only possible long-time asymptotic states of the system. These are analogous to the family of Oseen vortices in the 2D NSE. If one linearizes about one of these fixed points, one finds the spectrum of the linearized operator has a zero eigenvalue corresponding to motion along this manifold. The remainder of the spectrum lies strictly in the left half-plane. The next smallest eigenvalue is a simple eigenvalue $\lambda = -1/2$, with the rest of the spectrum having more negative real parts. Locally, invariant manifold theory allows one to construct a one-dimensional manifold tangent at the fixed point to the eigenfunction corresponding to this eigenvalue. Using the Cole-Hopf transformation, Beck and I extended this manifold globally and proved that these manifolds are normally stable - that is, if one enters a neighborhood of the manifold, one will remain in a neighborhood of the manifold for all subsequent times. We also showed that this manifold consists of exactly the diffusive N-waves previously identified as the important metastable states in Burgers equations by [KT01]. Thus, we referred to these manifolds as the "metastable manifolds." The final step in our construction was to show that "almost every" (in a sense made precise in [BW11b]) initial condition gives rise to a solution of Burgers equation which approaches one of these metastable manifolds on a short time scale. They evolve (due to the stability properties of the manifolds) slowly

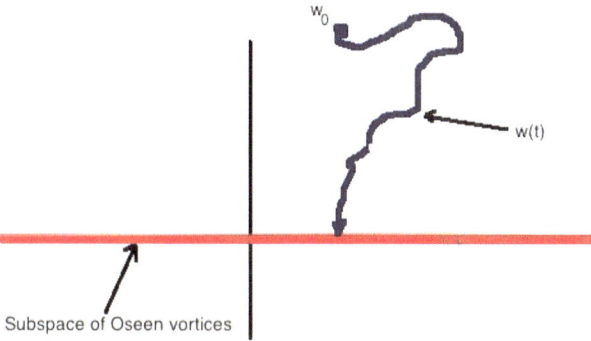

Figure 1.1 The phase space of the 2D NSE contains a line of fixed points (when expressed in terms of scaling variables) and almost every solution approaches some point on this line asymptotically, but we have little information about the rate of approach.

along the manifold until they eventually approach the long-time asymptotic state on the center manifold. If we represent this scenario graphically, we obtain the picture of the phase space of the weakly viscous Burgers equation: illustrated in Figure 1.2.

Comparing this with Figure 1.1, we see that in this case we have a much more detailed picture of the phase space structures which organize both the long-term and intermediate asymptotics, and we would now like to extend as much as possible of this model to understand the appearance of metastable states in the 2D NSE.

For the NSE equation on the torus we have already remarked that the only long-term asymptotic state is the zero solution. If we linearize the vorticity equation (1.42) around the zero solution we find, just as before, the heat equation,

$$\partial_t w = \nu \Delta w ,\qquad (1.46)$$

but this time with periodic boundary conditions. Note that unlike the case in $\mathbb{R}^2$ studied in the previous section, on the torus the heat equation has discrete spectrum (and a spectral gap) so there is no need to introduce scaling variables as we did there. Since we are only considering solutions of zero mean (see Remark 9), the eigenvalues of the right-hand side of (1.46) are

$$\lambda(m, \ell) = -\nu(m^2 + \delta^{-2}\ell^2) ,\quad (m, n) \neq (0, 0) ,\qquad (1.47)$$

with the corresponding eigenfunctions given by simple combinations of sines and cosines. We recall that the parameter δ measures the asymmetry of our domain, and if $\delta < 1$, then the smallest eigenvalue is given by $\lambda(1, 0) = -\nu$, with eigenfunction $w_{1,0}(x_1, x_2) = A \sin x_1$. More generally, we could choose the eigenfunction to be $A \sin x_1 + B \cos x_1$, but by a translation of the origin we can choose it to be proportional to $\sin x_1$. One expects on the basis of the general theory of dynamical systems that, modulo the technical difficulties that come from working in an infinite

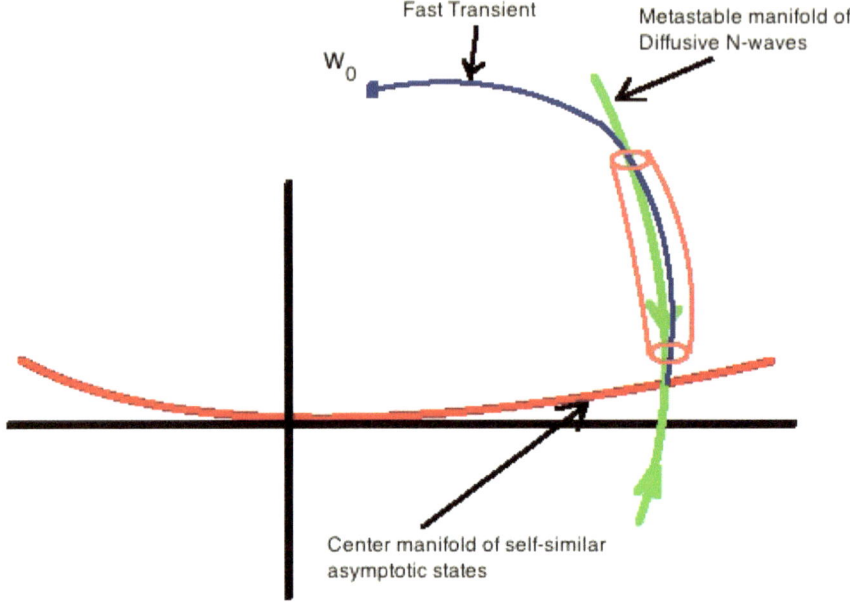

Figure 1.2 The phase space of the weakly viscous Burgers equation, showing the metastable manifolds of diffusive N-waves which govern the intermediate asymptotics.

dimensional phase space, one should be able to construct an invariant manifold for the semiflow generated by the NSE that is tangent at the origin to the eigenspace of the eigenvalue $\lambda(1,0) = -\nu$. However, in general, we will only be able to approximate this manifold in a small neighborhood of the fixed point $w \equiv 0$. In the case of Burgers equation we used the Cole-Hopf transformation to extend this local manifold globally, but that tool is not available here. Remarkably though, one can write down an invariant family of solutions of the full vorticity equation that is tangent at the origin to $A \sin x_1$. It is simply,

$$\omega^b(x_1, y_x) = Ae^{-\nu t} \sin(x_1) , \quad \mathbf{u}^b(x_1, x_2) = -Ae^{-\nu t} \begin{pmatrix} 0 \\ \cos x_1 \end{pmatrix} . \tag{1.48}$$

We'll refer to this family of states as bar states following the terminology of [YMC03], though these states are also known as Kolmogorov flows, and physically they represent a simple shear flow.

Remark 10. *There are a number of related explicit solutions of the 2D NSE. One can of course replace the sine functions in (1.48) with cosines, or take a linear combination of sine and cosine. There are also analogous states associated with the eigenvalues $-\nu m^2$ which are proportional to $\exp(-\nu m^2 t) \sin m x_1$, as well as solutions associated with the eigenvalues $-\nu(\ell/\delta)^2$ corresponding*

to shear flows oriented along the x_2 coordinate direction and proportional to
$\exp(-\nu(\ell/\delta)^2 t) \sin \ell x_2/\delta$. *More generally, if one takes any solutions of the heat*
equation with periodic boundary conditions which only depends on the variable
x_1, *this will give a solution of the two-dimensional vorticity equation, because if one*
checks the form of the velocity field given by the Biot-Savart law (1.45), one finds
that the nonlinear term in the equation vanishes identically.

Remark 11. *There are also explicit solutions analogous to the dipole solutions*
observed in [YMC03]. These appear in square domains (i.e., when the parameter
$\delta = 1$), *are often known as Taylor-Green vortices and they are solutions of the*
vorticity equation with

$$\omega(x_1, x_2.t) = Ae^{-\nu t}(\cos(x_1) + \cos(x_2)) \,, \quad \mathbf{u}(x_1, x_2, t) = Ae^{-\nu t}(-\sin(x_2), \sin(x_1)) \tag{1.49}$$

While we believe that the framework we use to discuss metastability of the bar states
below is probably also applicable to the dipole states, mathematically the analysis
is significantly harder, so we focus here just on the bar states.

Remark 12. *If one plots the constant vorticity contours of the bar states and the*
Taylor-Green vortices, one sees that they are very similar to those of the bar and
dipole states observed numerically.

However, there are some discrepancies. If, for example, one computes the stream
function associated with the bar states by solving Poisson's equation,

$$-\Delta \psi^b = \omega^b \,, \tag{1.50}$$

one sees that $\psi^b(x_1, x_2, t) = Ae^{-\nu t} \sin(x_1) = \omega^b(x_1, x_2, t)$ - i.e., the stream function
is a linear function of the vorticity. Plots of ψ vs. ω for numerical solutions of the
2D NSE (see Fig. 9 of [YMC03], for example) show that while for small values
of the vorticity the dependence is nearly linear, there is some departure from this
linear behavior at large values of the vorticity. Nonetheless we believe that the
bar states are good candidates for the metastable states in these systems because
once the system gets close to such a state (as it appears to do in the numerics)
the stability results described below show that it will remain nearby and actually
converge toward these states at a rate much faster than expected from viscous effects
alone.

We now examine the stability of the family of bar states. We hope both to show
that they attract nearby trajectories, and also to understand why they appear on
a time scale so much shorter than the viscous time scale. In the case of Burgers
equation we proved that the metastable manifold was normally stable with the aid
of the Cole-Hopf transformation. That tool is no longer available to us, so we resort
to a more direct, dynamical systems, approach, namely we linearize the NSE about
the bar states and study the evolution of this linearized equation. Linearizing (1.42)

about ω^b leads to the linear PDE

$$\partial_t w = \nu \Delta w - \mathbf{u}^b \cdot w - \mathbf{v} \cdot \nabla \omega^b \, , \tag{1.51}$$

where $\mathbf{v}$ is the velocity field associated with w. Because of the form of the velocity field $\mathbf{u}^b$ (see (1.48)), $\mathbf{u}^b \cdot w = Ae^{-\nu t} \sin(x_1)\partial_{x_2} w$. Likewise, since ω^b is independent of x_2, the last term in (1.51) also simplifies to $\mathbf{v}_1 \partial_{x_1} \omega^b = Ae^{-\nu t} \cos(x_1)\mathbf{v}_1$. From the Biot-Savart law we see that we can write $\mathbf{v}_1 = \partial_{x_2}(\Delta^{-1}w)$, where Δ^{-1} can be computed via its action on the Fourier series of w, i.e. $\widehat{\Delta^{-1}w}(m, \ell) = -\hat{w}(m, \ell)/(m^2 + (\ell/\delta)^2)$.

Thus, the last two terms on the RHS of (1.51) simplify and we are left with the linear equation

$$\partial_t w = \nu \Delta w - Ae^{-\nu t}(\cos(x_1))\partial_{x_2}(\mathbb{1} + \Delta^{-1})w \, . \tag{1.52}$$

Analyzing the stability of the family of bar states is more complicated than analyzing the stability of a fixed point of the equation because the linear equation (1.52) is non-autonomous. Nonetheless, computing the spectrum of the RHS of (1.52) for some fixed time t may give insight into the behavior of solutions. If we fix the time t and set $\tilde{A} = Ae^{-\nu t}$, we can compute the eigenvalues of

$$\mathcal{L}_{\nu,\tilde{A}} w = \nu \Delta w - \tilde{A}(\cos(x_1))\partial_{x_2}(\mathbb{1} + \Delta^{-1})w \, , \tag{1.53}$$

on the space of functions satisfying periodic boundary conditions. This is a relatively easy computation (numerically) if we express w as a Fourier series and consider the way $\mathcal{L}_{\nu,\tilde{A}}$ acts on these series. If $\hat{w}(m, \ell)$ are the Fourier coefficients of w, defined as in (1.44) we see that $\mathcal{L}_{\nu,\tilde{A}}$ doesn't "mix" different values of ℓ because there is no nontrivial x_2 dependence in the operator. Thus, we can consider separately the action of $\mathcal{L}_{\nu,\tilde{A}}$ on spaces of functions with different, fixed values of ℓ. Denoting this operator by $\hat{\mathcal{L}}^\ell_{\nu,\tilde{A}}$ we have

$$(\hat{\mathcal{L}}^\ell_{\nu,\tilde{A}}\hat{w})(m, \ell) = -\nu(m^2 + (\ell/\delta)^2) + \tag{1.54}$$

$$-i\frac{\tilde{A}\ell}{2\delta}\left[(1 - \frac{1}{(m-1)^2 + (\ell/\delta)^2})\hat{w}(m-1, \ell)\right.$$

$$\left. +(1 - \frac{1}{(m+1)^2 + (\ell/\delta)^2})\hat{w}(m+1, \ell)\right] \, .$$

Note that the operator $\hat{\mathcal{L}}^\ell_{\nu,\tilde{A}}$ has a special form. It has a real diagonal (and hence symmetric) piece with negative eigenvalues and a small coefficient in front of it, and a large off-diagonal piece which is "almost" skew-symmetric (due to the "i" in front of that term). In fact, as explained in [BW11a], a simple change of variables allows one to rewrite $\hat{\mathcal{L}}^\ell_{\nu,\tilde{A}}$ as the sum of a diagonal piece and an exactly skew-symmetric off-diagonal piece. As we'll see in the remainder of this section, such operators which arise frequently in fluid mechanics often have very special spectral properties.

As a first, simple remark about the properties of the operator $\hat{\mathcal{L}}^\ell_{v,\tilde{A}}$, note that if we are given any matrix of the form

$$L = D + A \tag{1.55}$$

where D is a real diagonal operator with eigenvalues lying in the set $\sigma_0 = \{\lambda \in \mathbb{R} \mid \lambda \le -v\}$, and with A a skew-symmetric matrix, then no matter how large A is (i.e., no matter how large its norm), the eigenvalues of L remain to the left (in the complex plane) of the line $\Re(\lambda) = -v$. This follows from the following simple calculation. Suppose that λ is an eigenvalue of L with normalized eigenvector v. Then we have

$$\lambda = \langle v, Lv \rangle = \langle v, (D + A)v \rangle = \langle (D - A)v, v \rangle$$
$$\overline{\lambda} = \langle Lv, v \rangle = \langle (D + A)v, v \rangle .$$

Adding these two expressions together and dividing by 2 we find

$$\Re(\lambda) = \langle v, Dv \rangle \le -v , \tag{1.56}$$

where the last inequality comes from the bound on the eigenvalues of D and the fact that $\|v\| = 1$.

In physical terms this means somewhat surprisingly that if we have a small, stable, symmetric dissipative linear operator, and add to it a skew-symmetric piece, a situation which comes up frequently in fluid mechanics when we linearize about some nontrivial solution of the NSE, we cannot destabilize the system, no matter how large the skew-symmetric piece is. In fact, what may happen is that the skew-symmetric piece, even though its own eigenvalues all lie on the imaginary axis, serves to further stabilize the system.

One example where this seems to occur is in the linearization of the 2D NSE in the whole plane about the Oseen vortex solutions that were discussed in the previous section. In that case, the dissipative diagonal part again just comes from the Laplacian term in the vorticity equation, while the skew-symmetric piece comes from the linearization of the nonlinear terms about the vortex. The stability properties of this linearization were first investigated numerically by Prochazka and Pullin [PP95] who found that as the Reynold's number increased (which means that the skew-symmetric piece of the operator became larger and larger with respect to the symmetric piece) the real part of almost all the eigenvalues of the linearization became increasingly negative - i.e. the vortex becomes more and more stable. The few eigenvalues whose real parts don't become more negative are fixed, independent of the Reynold's number and correspond to special, exact, symmetric solutions of the NSE.

A proposal to theoretically explain this stability phenomenon was first advanced by Gallagher, Gallay, and Nier [GGN09] who linked this behavior to the hypercoercivity method developed by Villani [Vil09] and were able to prove rigorously that this "enhanced stability" occurs in a model problem. More recently, W. Deng [Den12, Den11] has extended this method to the actual linearization about an Oseen vortex, at least for modes with a sufficiently strong angular dependence (i.e., if one expands the perturbation in a Fourier series in polar coordinates of the form $w(r, \theta) = \sum_n \hat{w}_n(r) \exp(in\theta)$, then n is required to be sufficiently large.)

Beck and I propose that a similar mechanism is responsible for the metastable properties of the bar states. We begin with numerical evidence that this is the case, by computing the eigenvalues of the operator $\hat{\mathcal{L}}^\ell_{\nu,\tilde{A}}$ as $\nu \to 0$. We find that in this limit, in which the skew-symmetric operator is much larger than the symmetric part, the eigenvalues not only all have negative real part, but also the real parts are proportional not to ν, as the eigenvalues of the symmetric part are, but rather to $\sqrt{\nu}$. This indicates the presence of a new time scale in the problem

$$\tau_{meta} \sim \frac{1}{\sqrt{\nu}} << \tau_{viscous} \sim \frac{1}{\nu} \qquad (1.57)$$

which is much shorter than the viscous time scale when ν is small.

To illustrate this effect, we first start with a simple example in which we take a 41×41 matrix approximation to the operator $\hat{\mathcal{L}}^\ell_{\nu,\tilde{A}}$. We choose the parameters $A = 1$, $\nu = 0.01$, $\delta = 0.9$ and $\ell = 1$, though we would obtain very similar results for essentially any other choices of parameters. The three figures below show first of all the eigenvalues of the diagonal part of the operator, then the off-diagonal, nearly anti-symmetric part of the operator, and finally the eigenvalues of the full operator, illustrating how the interplay between these two pieces actually results in eigenvalues whose real part is more negative, than the operator without the anti-symmetric part (Figure 1.3).

To study more quantitatively the effects of this interplay, and in particular, to examine the dependence of the eigenvalues on the viscosity, ν, the next figure

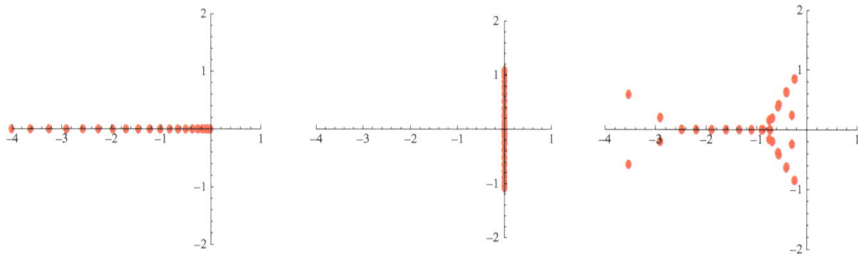

Figure 1.3 A numerical simulation showing how the interplay between the symmetric and anti-symmetric parts of an operator can lead to more negative eigenvalues (i.e., enhanced stability) than either part alone.

Figure 1.4 A log-log plot of ν against the absolute value of the real part of the least negative eigenvalue, superimposed on a line of slope $1/2$.

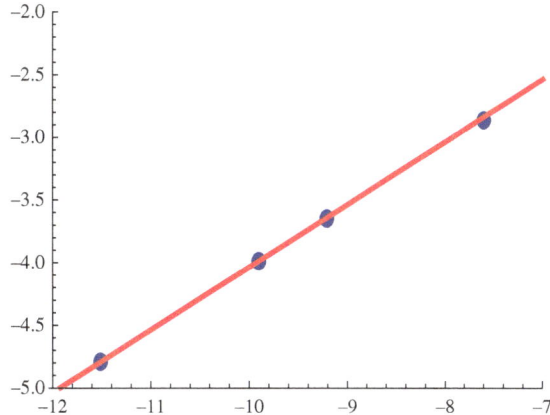

considers the least negative eigenvalue of a 201×201 matrix approximation to $\hat{\mathcal{L}}^{\ell}_{\nu,\tilde{A}}$. The parameters A, δ, and ℓ are as above, but now we choose four different values of $\nu = 0.0005, 0.0001, 0.00005, 0.00001$. We then plot the logarithm of ν against the logarithm of the absolute value of the real part of the least negative eigenvalue. Intuitively, this eigenvalue should determine the least stable mode when we perturb the bar states. Superimposed on the log-log plot is a line of slope $1/2$, which would correspond to the real part of the eigenvalue being proportional to $\sqrt{\nu}$. As one can see, the agreement is very good (Figure 1.4).

Remark 13. *There is one special case in which the eigenvalues of $\hat{\mathcal{L}}^{\ell}_{\nu,\tilde{A}}$ do scale as ν and that is when $\ell = 0$. These correspond to family of x_2 independent solutions of the vorticity equation discussed in Remark 10. There are also a small number of eigenvalues which scale anomalously when $\ell = \pm1$. They are discussed in detail in [BW11a]. While the origin of these "anomalous" states is not clear, it may be that they are somehow connected with the dipole states.*

The eigenvalue calculations above suggest that solutions of (1.53) will decay with rate $\sim \sqrt{\nu}$, but for highly non-symmetric, non-autonomous operators like $\mathcal{L}$, it is well known that one cannot automatically assume that the spectrum of the linearized operator at a particular time determines the decay rate of solutions. Often, a more relevant quantity is the pseudo-spectrum, [TE05]. The behavior of the pseudo-spectrum may be qualitatively different than that of the eigenvalues and in the model of the linearization of the NSE about the Oseen vortex investigated by Gallagher et al [GGN09], this turns out to be the case. However, for our problem numerical calculations suggest that the pseudospectral bounds also scale like $\sqrt{\nu}$, so these computations lead to conclusions about the stability of the bar states similar to those obtained from the eigenvalue calculations above.

Because of the difficulty in obtaining information about the decay rates of the evolution generated by $\hat{\mathcal{L}}^{\ell}_{\nu,\tilde{A}}$ from either the spectrum or pseudo-spectrum of the operator at any fixed time, we will instead study the evolution with the aid of the hypercoercivity method developed by Villani [Vil09]. This method was developed to deal specifically with evolutions generated by operators of the form $L = A^*A + B$, where B is skew-symmetric. The method can account for the enhanced stability of the sort we observe by exploiting the lack of commutativity between A and B. This point is crucial - if $[A, B] = 0$, we could simultaneously diagonalize the two operators and wouldn't observe a qualitative difference in the behavior of the real parts of the eigenvalues of A^*A and $A^*A + B$. Villani systematized this by considering the time decay of a special functional which incorporates this non-commutativity and then relating this functional to more standard norms of solutions.

At the moment we're unable to apply Villani's method to the full linearized equation (1.53) due to the presence of the non-local term $\cos(x_1)(\Delta^{-1}w)$. We have computed the spectrum of the operator $\hat{\mathcal{L}}^\ell_{\nu,\tilde{A}}$ both with this term present (those were the computations presented above) and without this term (see Figure 1.5, just below) and the spectrum is very similar in both cases. In particular, the real parts of the eigenvalues display the same $\sqrt{\nu}$ scaling. In addition the non-local term is a compact, lower order perturbation so we believe it will have a small influence on the evolution of the system. Thus, instead of studying the evolution generated by $\hat{\mathcal{L}}^\ell_{\nu,\tilde{A}}$, we consider the approximate equation

$$\partial_t w = \mathcal{L}^\ell_{approx} w = \nu \Delta w - Ae^{-\nu t}(\cos(x_1))\partial_{x_2} w , \qquad (1.58)$$

or, if we again expand w in terms of its Fourier series and consider terms with the same fixed value of ℓ, as we did in (1.54), we have

$$\partial_t w = \mathcal{L}^\ell_{approx} w = -\nu(\partial^2_{x_1} - \ell^2)w + i\frac{\ell}{8}Ae^{-\nu t}(\cos(x_1))w . \qquad (1.59)$$

Remark 14. *Note that one other advantage of (1.59) as compared to (1.52) is that the operator on the RHS of (1.59) is already a sum of a symmetric piece plus a skew-*

Figure 1.5 A plot of the eigenvalues of the approximate linear operator in (1.58). The parameter values are the same as in Figure 1.3

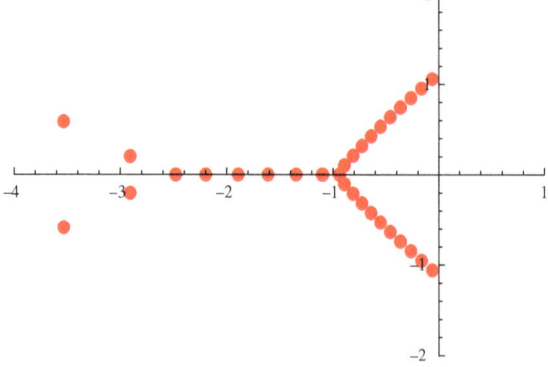

symmetric piece, without resorting the initial change of variables needed to bring $\hat{\mathcal{L}}^{\ell}_{\nu,\tilde{A}}$ *into this form.*

Remark 15. *In the study of the linearization of the 2D NSE about the Oseen vortex a similar nonlocal term appeared in linearization. In [Den12], Deng studied the pseudo-spectrum of the full linearization of the equation and the approximation to the linearization obtained by dropping the nonlocal term. She found similar results in both cases.*

Remark 16. *Note that (1.59) no longer contains any explicit dependence in x_2. Thus, for the remainder of this section we will replace x_1 by x to simplify the notation.*

As mentioned above, Villani's method exploits the non-commutativity of the symmetric and skew-symmetric parts of the operator and with that in mind we define the following operators

$$B^{\ell}w = i\frac{\ell}{\delta}Ae^{-\nu t}(\cos(x))w \, , \quad C^{\ell}w = [\partial_{x_1}, B^{\ell}]w = i\frac{\ell}{\delta}Ae^{-\nu t}(\sin(x))w \qquad (1.60)$$

Note that like B^{ℓ}, C^{ℓ} is also skew-symmetric, B^{ℓ} and C^{ℓ} commute with each other, and both are bounded operators.

Following Villani, we now define a functional that incorporates the effects of C^{ℓ} on the evolution, namely,

$$\Phi^{\ell}(t) = \|w\|^2 + \alpha\|\partial_x w\|^2 - 2\beta\Re(\partial_x w, C^{\ell}w) + \gamma(C^{\ell}wl, C^{\ell}w). \qquad (1.61)$$

where the constants α, β, and γ will be chosen in the course of the proof. The first restriction we place on their values comes from the fact that we want Φ^{ℓ} to control some norm of the solution of (1.59). If we require that

$$\beta^2 < \alpha\gamma/4 \, , \qquad (1.62)$$

then we have

$$\|w\|^2 + \frac{\alpha}{2}\|w_x\|^2 + \frac{\gamma}{2}\|C^{\ell}w\|^2 < \Phi^{\ell}(t) < \|w\|^2 + \frac{3\alpha}{2}\|w_x\|^2 + \frac{3\gamma}{2}\|C^{\ell}w\|^2.$$

Now consider the time rate of change of the Φ^{ℓ} along a solution of (1.59). This is almost identical to the analogous computation in [Vil09], except for minor modifications due to the fact that the operators B^{ℓ} and C^{ℓ} are non-autonomous in our case. Thus, we obtain:

$$\frac{d}{dt}\Phi^{\ell}(t) = ((w_t, w) + (w, w_t)) + \alpha\left((\partial_x w_t, \partial_x w) + (\partial_x w, \partial_x w_t)\right) \qquad (1.63)$$

$$-2\beta\Re\left((\partial_x w_t, C^{\ell}w) + (\partial_x w, C^{\ell}w_t)\right) + \gamma\left((C^{\ell}w_t, C^{\ell}w) + (C^{\ell}w, C^{\ell}w_t)\right)$$

$$-2\beta\Re(\partial_x w, \frac{dC^\ell}{dt}w) + \gamma\left((\frac{dC^\ell}{dt}w, C^\ell w) + (C^\ell w, \frac{dC^\ell}{dt}w)\right).$$

We now use (1.59) and the properties of B^ℓ and C^ℓ (specifically their anti-symmetry) to control each of the terms in (1.63). The details of that procedure are set out in [BW11a] - in this review we just focus on a few representative terms, including those that give the enhanced stability, to explain the ideas involved.

For instance, consider the terms

$$(w_t, w) + (w, w_t) = -\nu\left[(\partial_x^2 w, w) + (w, \partial_x^2 w)\right] - 2\nu\ell^2(w, w)$$
$$+(B^\ell w, w) + (w, B^\ell w) \qquad (1.64)$$
$$= -2\nu\left[\|w\|_{L^2}^2 + \|w_x\|_{L^2}^2\right],$$

where the terms involving B^ℓ vanish due to anti-symmetry. Note that the remaining terms contribute to the decay of Φ^ℓ, but only with a rate proportional to ν - so this would represent decay on the viscous time scale.

To see where the accelerated decay comes from, we consider those terms proportional to β, since they are the ones which exploit the lack of commutativity between the symmetric and anti-symmetric parts of $\mathcal{L}_{approx}^\ell$. Thus,

$$(\partial_x w_t, C^\ell w) + (\partial_x w, C^\ell w_t) = -2\ell^2\nu\Re(w_x.C^\ell w) + \nu\left[(w_{xxx}, C^\ell w) + (w_x, C^\ell w_{xx})\right]$$
$$+(\partial_x(B^\ell w), C^\ell w) + (w_x, C^\ell(B^\ell w)). \qquad (1.65)$$

The terms involving derivatives of w can be handled with the aid of integration by parts and then absorbed in other terms because they are all proportional to ν. The two remaining terms can be combined if we recall from (1.60) that $\partial_x B^\ell = C^\ell + B^\ell\partial_x$, while $C^\ell B^\ell = B^\ell C^\ell$. Thus, we find

$$(\partial_x(B^\ell w), C^\ell w) + (w_x, C^\ell(B^\ell w)) = (C^\ell w, C^\ell w) + (B^\ell w_x, C^\ell w) + (w_x, B^\ell(C^\ell w))$$
$$= \|C^\ell w\|^2, \qquad (1.66)$$

again, using the anti-symmetry of B^ℓ. Taking into account the fact that this term has a negative coefficient in front of it, we see that it gives a large (i.e., $\mathcal{O}(1)$, rather than $\mathcal{O}(\nu)$) negative contribution to $d\Phi^\ell/dt$ and is responsible for the accelerated convergence rate.

There are still a number of problems which must be overcome. The remaining terms must all be carefully bounded, and in particular, while the term discussed in the preceding inequality will yield an accelerated, convergence rate for the part of Φ^ℓ proportional to $\|C^\ell w\|^2$, one must show that it can also yield a bound for the other terms in Φ^ℓ. The details of these estimates are provided in [BW11a], where one finds that there exist positive constants M and K, independent of ν, such that

$$\Phi^{\ell}(t) \leq K e^{-M\sqrt{\nu} t} \Phi^{\ell}(0) \; . \tag{1.67}$$

It is also explained there how one can combine the estimates for different values of ℓ to show that the norm of the full solution of (1.58) also decay with a rate proportional to $\sqrt{\nu}$.

Thus we see that the interplay between the symmetric and anti-symmetric parts of the linearized operator leads to a decay rate much faster than the $\mathcal{O}(\nu)$ viscous decay rate, and consequently a time scale associated with the convergence to the bar states that is much faster than suggested by viscous effects alone. For instance, for a Reynold's number of $\sim 10^4$, (in the non-dimensionalized variables in which we are working, this would be $\nu \sim 10^{-4}$), these estimates suggest an approach toward the metastable solutions like bar states on a time scale of the order of $\tau_{meta} \sim 100$ time units, rather than the $10{,}000$ time units expected if the convergence were governed by the viscous time scale. We note that these convergence rates and Reynold's numbers are qualitatively similar to those observed in numerical simulations of these systems.

Of course the picture in the case of metastable behavior for the 2D NSE is still far less complete than for Burgers equation. The most important missing piece is the fact that so far, the rigorous estimates only apply to the approximation of the true linear evolution obtained by dropping the nonlocal term in the equation. However, as we have seen from the numerical computations presented above, the spectral properties of the operator with or without the nonlocal term are very similar, and we hope that the sort of techniques Deng developed in [Den11] will also apply here and allow us to treat the full linearized operator.

A second important difference between Burgers equation and the 2D NSE is that in Burgers equation the viscous N-waves are the only metastable states for almost all solutions, but for the NSE there are other candidates, such as the dipole states observed in numerics (see Remark 11), or the states discussed in Remark 10 which like the bar states only depend on one of the two spatial variables. This means that the final picture of metastability for the NSE will undoubtedly be more complicated than the simple picture in Figure 1.2 for Burgers equation. Nonetheless, we feel that these dynamical systems ideas can yield important simplifying insights into the intermediate asymptotics of this important physical system and can explain how time scales much shorter than the viscous time scales can arise from the operators which have a small, symmetric dissipative piece, and a large, anti-symmetric component, which arise frequently in the study of fluid motions.

4   Finite Dimensional Attractors for the Navier-Stokes equations

In the two previous sections we focussed on the behavior of the two-dimensional Navier-Stokes equations in the absence of forcing and studied the types of structures that emerged on intermediate and long time scales. In the present section we consider solutions of the equation when it is subjected to external forcing. In practice, such forcing might arise through shearing at the boundaries of the domain, as is the case in the Taylor-Couette experiment, but as is often done in theoretical studies we will allow for a general body force, i.e. we will consider the form of the equations in (1.8) and (1.9).

$$\partial_t \mathbf{u} + \mathbf{u} \cdot \nabla \mathbf{u} = \nu \Delta \mathbf{u} - \nabla p + \mathbf{g} \tag{1.68}$$

$$\nabla \cdot \mathbf{u} = 0 . \tag{1.69}$$

Here the forcing term $\mathbf{g}$ is some general vector valued function whose smoothness we specify below.

Numerical studies show that even in the presence of certain types of forcing, the structures we studied in previous sections may still play a dominant role in the dynamical behavior of the system [BS09]. However, in this section we focus on more general properties of the system such as the existence and dimension of the attractor for the system. We will present fewer details of the calculations in this section than in the prior ones since the work described here is somewhat older than that in the previous sections and several excellent monograph length reviews of these results already exist [Rob01, DG95].

While most investigations of these questions work directly with the fluid velocity, in order to keep the discussion as close as possible to prior sections we will continue to work with the vorticity formulation of the problem. This has the disadvantage that we are restricted essentially to domains without boundary and thus we focus here as in section three on a rectangular domain with periodic boundary conditions, so we consider functions in

$$L^2(\mathbb{T}_\delta^2) = \{u_j \in L^2 \mid u_j(x_1, x_2) = u_j(x_1 + 2\pi, x_2) = u_j(x_1, x_2 + 2\pi\delta) , \, j = 1, 2\} . \tag{1.70}$$

Assuming that the forcing $\mathbf{g}$ is differentiable, we can then take the curl of (1.68) to obtain the scalar equation

$$\partial_t \omega = \nu \Delta \omega - \mathbf{u} \cdot \nabla \omega + f , \tag{1.71}$$

where $\mathbf{u}$ is again the velocity field recovered from the vorticity via the Biot-Savart law (1.45), and $f = \partial_{x_1} \mathbf{g}_2 - \partial_{x_2} \mathbf{g}_1$.

Proving the existence and uniqueness of solutions for this equation is straightforward so we will assume that has been done and focus instead on more detailed aspects of its dynamics. Recall from the introduction that for an infinite dimensional dynamical system, constructing an attractor requires us to find a pre-compact absorbing ball. That follows in the present circumstances from straightforward energy estimates. If we multiply (1.71) by ω and integrate over the torus, we find that

$$\partial_t \|\omega(t)\|_{L^2} = -\nu \|\nabla\omega\|_{L^2}^2 - \int_{\mathbb{T}_\delta^2} \omega(\mathbf{u}\cdot\nabla\omega)dx + \int_{\mathbb{T}_\delta^2} \omega f dx \,. \tag{1.72}$$

Consider each of the terms on the RHS of this equation in turn. Since $\int \omega dx = 0$, we can apply Poincaré's inequality to the first term and bound it by $-c_p \nu \|\omega\|_{L^2}^2$. Turning to the second term, the incompressibility of the velocity field $\mathbf{u}$ allows us to rewrite it as:

$$\int_{\mathbb{T}_\delta^2} \omega(\mathbf{u}\cdot\nabla\omega)dx = \frac{1}{2}\int_{\mathbb{T}_\delta^2} \mathbf{u}\cdot\nabla\omega^2 dx = \frac{1}{2}\int_{\mathbb{T}_\delta^2} \nabla\cdot(\mathbf{u}\omega^2)dx = 0 \,, \tag{1.73}$$

where the last inequality follows from the periodic boundary conditions. Finally the Cauchy-Schwartz inequality implies that $|\int \omega f dx| \leq \|\omega\|_{L^2}\|f\|_{L^2} \leq \frac{1}{2}c_p \nu \|\omega\|_{L^2}^2 + \frac{1}{2c_p \nu}\|f\|_{L^2}^2$. Inserting these estimates into (1.72) we find

$$\partial_t \|\omega(t)\|_{L^2} = -\frac{1}{2}c_p \nu \|\omega\|_{L^2}^2 + \frac{1}{2c_p \nu}\|f\|_{L^2}^2 \,. \tag{1.74}$$

If we now assume that the forcing function is an element of $L^\infty([0,\infty); L^2(\mathbb{T}_\delta^2))$, we can apply Gronwall's inequality to conclude that

$$\|\omega(t)\|_{L^2}^2 \leq e^{-c_p \nu t/2}\|\omega_0\|_{L^2}^2 + \frac{2}{c_p \nu}(1 - e^{-c_p \nu t/2})\|f\|_{L^\infty([0,\infty);L^2)} \,. \tag{1.75}$$

and so we see that all solutions of (1.71) will eventually enter the ball of radius $R = \frac{2}{c_p \nu}\|f\|_{L^\infty([0,\infty);L^2)}$.

As noted in the introduction, in infinite dimensional systems, the existence of an absorbing ball is not enough in itself to guarantee an attractor. We must show that the absorbing set is also (pre-)compact in L^2. One does this by showing the boundedness of the absorbing ball in H^s, with $s \geq 1$, and using the fact that H^s is compactly embedded in L^2. The boundedness of the H^s norm follows by energy estimates similar to those in (1.72)–(1.74). They are somewhat more complicated in this case though because the nonlinear term no longer vanishes as in (1.73). One still obtains a bound on the H^s norm though and as a consequence the existence of a bounded, compact attractor for the forced two-dimensional NSE. In fact, in two-dimensions, there is a whole sequence of energy estimates linking and bounding the various Sobelev norms of the solution - for an explication of these estimates, sometimes known as the ladder theorems, see [DG95].

As discussed in the introduction, one advantage of the existence of an attractor is that it often allows one to focus attention on a much smaller dimensional system than the original, infinite dimensional PDE. In fact, the attractor for the two-dimensional NSE is finite dimensional and one can estimate its dimension in terms of the forcing function f. This fact was first established for a class of reaction-diffusion equations by Mallet-Paret [MP76]. In the context of the two-dimensional NSE is first studied by Constantin and Foias, [CF85]. It is well known that the attractor of a chaotic dynamical system is often a fractal set and so to measure it, we must use a *fractal* dimension. There are several common fractal dimensions including the Hausdorff dimension and the capacity dimension. Estimates of the both these quantities exist in the case of the attractor in the two-dimensional NSE, but we'll focus on the capacity here because it is the simplest to define. Given a set $\mathcal{A}$ in the phase-space of our PDE, define $N(r) = $ min number of balls of radius r needed to cover $\mathcal{A}$. Then the capacity dimension of $\mathcal{A}$ is defined to be:

$$d_C(\mathcal{A}) = \limsup_{r \to 0} \frac{N(r)}{\log(1/r)} . \qquad (1.76)$$

It is easy to check that if one takes a segment of a curve or surface in $\mathbb{R}^d$, the capacity dimension gives the usual and expected values of 1 and 2, respectively. Suppose now that we consider the attractor for the two-dimensional NSE. The basic idea relating the dimensional estimates to the dynamics of the system is the following. Suppose one covers the attractor with d-dimensional balls of radius r. One can show that if the attractor is finite dimensional it suffices to use finite (although perhaps large) dimensional balls to compute the dimension. If one considers how these balls will be distorted by the flow, say by the time-one map of the system, they will typically be elongated in some directions, corresponding to the chaotic stretching that occurs in the attractor. However, the high-frequency modes in the Navier-Stokes equation are very strongly damped, and if we choose d, the dimension of the covering balls, sufficiently large, the number of contracting directions will overwhelm the effect of the expanding directions and result in the total volume of the ball shrinking. The capacity is defined in terms of the limit as the radius of the balls tends to zero so we are most interested in the action of the flow on balls of very small radius. In this limit, the action of the flow is well approximated by the linearization of the flow, which will distort our original ball into a d-dimensional ellipsoid. The lengths of the axes of this ellipsoid can then be related to the radius of the original ball by the singular values of the linearized system, which are in turn related to the Lyapunov exponents of the dynamical system. If the Lyapunov exponents of the system are $\mu_1 \geq \mu_2 \geq \ldots$, (where we have listed the exponents along with their multiplicity), then the shortest axis of the deformed ellipsoid should have length $\sim r e^{\mu_d}$, and the volume on the ellipsoid should be approximately $\sim r^d e^{\mu_1 + \cdots \mu_d}$. This in turn gives a recursion relation between $N(r)$ and $N(e^{\mu_d})$, which can then be used to evaluate the lim sup in (1.76). The details of this calculation are very clearly explained in ([DG95], p. 67) and lead to an estimate of the dimension in terms of the Lyapunov exponents.[2]

There are a number of points that must be addressed in order to make this argument rigorous including:

---

[2]The link between the attractor dimension and the Lyapunov exponents was first proposed by Kaplan and Yorke, [KY79], and is often known as the Kaplan-Yorke formula.

- The Lyapunov exponents concern the asymptotic growth rates of the linearized flow, not the amount of growth or shrinking that occurs in the time-one map.
- Typically, the Lyapunov exponents will depend on the initial point we choose, so it's not clear why this argument leads to a uniform estimate on the dimension of the attractor.
- Finally, we would like to have an estimate on the attractor dimension in terms of accessible quantities in the equation, like the viscosity, v, or the forcing f, rather than generally unknown Lyapunov exponents.

Constantin and Foias address the first two points by introducing global Lyapunov exponents. They consider the usual definition of the Lyapunov exponents in terms of the growth rates of n-dimensional volumes induced by the linearized flow, but before taking the infinite time limit of these growth rates, they take the supremum of the growth rates over the attractor. This gives an estimate of the growth rate independent of the starting point, but might lead to an overestimate of the dimension. They also give rigorous estimates relating the size of a ball of radius r, transported for a finite time t, to these global Lyapunov exponents, as well of making rigorous the details of the Kaplan-Yorke formula for infinite dimensional systems like the NSE.

To relate the Lyapunov exponents of the NSE to the quantities that appear in the equation, note that if we linearize (1.71) about a solution $\omega(t)$, with associated velocity field $\mathbf{u}(t)$, the linearized equation takes the form:

$$\partial_t W = v \Delta W - \mathbf{u} \cdot \nabla W - \mathbf{U} \cdot \nabla \omega , \tag{1.77}$$

where $\mathbf{U}$ is the velocity field associated with W.

Recall that the estimates of dimension were based on how the linearized flow causes d-dimensional spheres to expand or contract. Thus, instead of studying (1.77) directly, we consider the evolution of $(P_d W)$, the projection of W onto an d-dimensional subspace of the infinite dimensional phase space, and consider the maximum growth rate over all such projections. The growth rate of such projections can be controlled by energy estimates of the type that lead to (1.75), and one finds

$$d_C \leq C \left(\frac{\|\mathbf{g}\|_{L^2}}{v^2} \right)^{2/3} \left(1 + \log \left(\frac{4\pi^2 \delta}{v^2} \|\mathbf{g}\|_{L^2} \right) \right) . \tag{1.78}$$

In this estimate, the factor of $4\pi^2\delta$ is just the area of the domain on which the equation evolves, and $\mathbf{g}$ is the forcing term in the equation for the velocity field, not the vorticity, and this force is assumed to be constant in time. Note that although the forcing function for the fluid velocity appears in the estimate of dimension, the actual estimates are done in the vorticity formulation which seems to yield sharper bounds on the attractor dimension than working directly with the velocity [CFT88]. For more details on the derivation of these estimates, one can consult either the original paper, [CF85], which explains the ideas behind the proof very clearly, or the monograph, [DG95].

Remark 17. *There is a non-rigorous estimate of the dimension of the attractor of the two-dimensional Navier-Stoke equation based on a scaling theory of turbulence. Remarkably, the rigorous estimate (1.78) agrees with the prediction of the scaling theory, up to the factor of* $\left(1 + \log\left(\frac{4\pi^2\delta}{\nu^2}\|\mathbf{g}\|_{L^2}\right)\right)$.

Remark 18. *Bounding the dimension of the attractor is one way to quantify the essentially finite dimensional nature of the two-dimensional NSE semi-flow. Another approach focuses on showing that there is a finite number of* determining modes *or* determining nodes. *In this case, one asks whether or not the knowledge of the behavior of the projection of the solution on a finite number of modes, or its value at a finite number of points, is sufficient to determine its long time behavior. More precisely, let P_d be the projection onto an d-dimensional subspace discussed above. One says that the system has d determining modes if for some choice of P_d, any two solutions $\omega(t)$ and $\tilde{\omega}(t)$ for which*

$$\lim_{t\to\infty}\|P_d(\omega(t)) - P_d(\tilde{\omega}(t))\|_{L^2} = 0, \tag{1.79}$$

must also have

$$\lim_{t\to\infty}\|\omega(t) - \tilde{\omega}(t)\|_{L^2} = 0, \tag{1.80}$$

with a similar definition for determining nodes. Note that this definition does not imply that given a knowledge of the first d modes of the system we can reconstruct the remaining modes, but merely that those (infinitely many) other modes cannot change the long-term asymptotics of the solution.

The first estimates establishing that the solutions of the two-dimensional NSE had a finite number of determining modes were by Foias and Prodi [FP67]. More recently, Jones and Titi [JT93] have shown that for periodic boundary conditions, the number of determining modes is bounded by

$$d \leq C\frac{\|\mathbf{g}\|_{L^2}}{\nu^2}, \tag{1.81}$$

which is larger than estimate for the dimension of the attractor established above.

While the finite dimensionality of the attractor for the two-dimensional NSE implies that the asymptotic behavior is determined by the motions on some "small" piece of the phase space, it doesn't immediately imply that there is a smooth finite dimensional dynamical system whose motion reproduces the long-time behavior of the NSE. The reason is that one knows very little about the structure of the attractor and in fact, for chaotic systems, it is expected to be a complicated, fractal set. Thus, simply restricting the NSE to the attractor will not generally lead to a smooth, finite dimensional, dynamical system. An alternate approach which has been pursued is to attempt to construct a finite dimensional manifold which is:

- smooth, or at least Lipshitz,
- invariant with respect to the semi-flow defined by the NSE, and
- attracts any solution of the equation at an exponential rate.

Such manifolds, if they exist, are called *inertial manifolds*.

If a PDE possesses an inertial manifold, then by its invariance and attractivity, the attractor must lie inside it. Thus, by restricting the PDE to the inertial manifold one obtains a smooth (or at least Lipshitz), finite dimensional dynamical system which captures all the asymptotic behavior of the original, infinite dimensional system.

Inertial manifolds bear much resemblance to center manifolds in finite dimensional systems of ordinary differential equations. Recall that the existence theory for center manifolds requires that the "spectral gap," that is, the difference between the real parts of the stable and unstable eigenvalues and the center eigenvalues, must be large in comparison with the Lipschitz constant of the nonlinear term in the equation.

In the inertial manifold case, this becomes the requirement that there be a large gap between the eigenvalues corresponding to modes related to the inertial manifold and those "off" the manifold. Typically the distinction between these two sets of modes is expressed by writing the inertial manifold as a graph of a function whose domain is the span of the first set of these modes and whose range is the span of the modes off the manifold. The size of the gap is again related to the properties of the nonlinear term, but are more complicated than those in the ODE case because the nonlinear term in the PDE is often not Lipshitz on the whole phase space of the system. For example, consider the nonlinear term in the vorticity equation (1.71). The presence of the derivative in the nonlinearity means it is not even defined on the whole phase space, $L^2(\mathbb{T}_\delta^2)$, let alone Lipschitz. This lack of smoothness results in much more stringent spectral conditions than in the center manifold case.

For nonlinear terms of the sort that appear in the NSE, the best estimates to date for the spectral gap require that if the eigenvalues of the linear part of the equation are ordered as $0 \geq -\lambda_1 \geq -\lambda_2 \geq \ldots$, then proof of existence for the inertial manifold requires that for a manifold on dimension N, one must have:

$$\lambda_{N+1} - \lambda_N \geq K(\lambda_{N+1}^{1/2} + \lambda_N^{1/2}) . \tag{1.82}$$

For (1.71), the eigenvalues of the linear part can be explicitly computed as $\nu(m^2 + (\ell/\delta)^2)$, and one sees that the large gaps required by (1.82) do not exist. As a consequence, there is no proof to date that the two-dimensional NSE possesses an inertial manifold.

There are, however, large classes of dissipative PDEs for which the existence of an inertial manifold has been proven. Most of these fall into one of two categories:

1. 1-dimensional equations: In one dimension, if one considers a PDE with linear part

$$\partial_t u = \nu \partial_x^r u , \tag{1.83}$$

with periodic boundaries on the interval $[-\pi, \pi]$, the eigenvalues are $-\lambda_n = -\nu n^{2r}$. Thus, for $r > 1/2$, one obtains arbitrarily large gaps between eigenvalues as n grows. Using this growth, the existence of inertial manifolds for equations like the Kuramoto-Sivishinsky and Cahn-Hilliard, [FST85, FST88, CFNT89].

2. Equations with "nicer" nonlinear terms: If the nonlinear term in the equation is a smooth function on the phase space of the equation, as is true for many reaction-diffusion equations, the spectral gap condition can be greatly weakened and the existence of inertial manifolds for many two-dimensional domains, and even some special three-dimensional domains has been established, [MPS88].

In conclusion, we have seen that the tools of dynamical systems theory provide a number of insights into the qualitative behavior of solutions of dissipative PDEs, especially the two-dimensional Navier-Stokes equations. The identification of finite dimensional invariant structures in the infinite dimensional phase space of these systems, be they attractors, or invariant manifolds, allows one to better understand the long-time asymptotic behavior of both the forced and unforced Navier-Stokes equations, as well as providing a way of understanding the emergence of intermediate time scales in these systems.

Acknowledgements The support of the author's research by the National Science Foundation grants, DMS-0908093 and DMS-1311553 is gratefully acknowledged. For those parts of this survey which describe my own research it is a pleasure to thank my collaborators - Thierry Gallay and Margaret Beck for the theoretical results in Sections 2 and 3, and Alethea Barbaro, Ray Nagem, Guido Sandri, and David Uminsky for the numerical methods described in Section 2.

References

[BA94] Matania Ben-Artzi. Global solutions of two-dimensional Navier-Stokes and Euler equations. *Arch. Rational Mech. Anal.*, 128(4):329–358, 1994.

[BJ89] Peter W. Bates and Christopher K. R. T. Jones. Invariant manifolds for semilinear partial differential equations. In *Dynamics reported, Vol. 2*, volume 2 of *Dynam. Report. Ser. Dynam. Systems Appl.*, pages 1–38. Wiley, Chichester, 1989.

[BS09] Freddy Bouchet and E. Simonnet. Random changes of flow topology in two-dimensional and geophysical turbulence. *Phys. Rev. Lett.*, 102(094504), 2009.

[BW11a] Margaret Beck and C. Eugene Wayne. Metastability and rapid convergence to quasi-stationary bar states for the 2D Navier-Stokes equations. Proc. Roy. Soc. Edin., Sec. A Math., 143(5):905–927, 2013.

[BW11b] Margaret Beck and C. Eugene Wayne. Using global invariant manifolds to understand metastability in the Burgers equation with small viscosity. *SIAM Review*, 53(1):129–153 [Expanded and revised version of paper of the same title published originally in SIAM J. Appl. Dyn. Syst. 8 (2009), no. 3, 1043–1065], 2011.

[CF85] P. Constantin and C. Foias. Global Lyapunov exponents, Kaplan-Yorke formulas and the dimension of the attractors for 2D Navier-Stokes equations. *Comm. Pure Appl. Math.*, 38(1):1–27, 1985.

[CFNT89] P. Constantin, C. Foias, B. Nicolaenko, and R. Temam. *Integral manifolds and inertial manifolds for dissipative partial differential equations*, volume 70 of *Applied Mathematical Sciences*. Springer-Verlag, New York, 1989.

[CFT88] P. Constantin, C. Foias, and R. Temam. On the dimension of the attractors in two-dimensional turbulence. *Phys. D*, 30(3):284–296, 1988.

[CHT97] Xu-Yan Chen, Jack K. Hale, and Bin Tan. Invariant foliations for C^1 semigroups in Banach spaces. *J. Differential Equations*, 139(2):283–318, 1997.

[CL95] Eric A. Carlen and Michael Loss. Optimal smoothing and decay estimates for viscously damped conservation laws, with applications to the 2-D Navier-Stokes equation. *Duke Math. J.*, 81(1):135–157 (1996), 1995. A celebration of John F. Nash, Jr.

[Den11] Wen Deng. Pseudospectrum for Oseen vortices operators. IMRN, 2013(9):1985–1999, 2013.

[Den12] Wen Deng. *Etude du pseudo-spectre d'opérateurs non auto-adjoints liés à la mécanique des fluides.* PhD thesis, Université Pierre et Marie Curie, 2012.

[DG95] Charles R. Doering and J. D. Gibbon. *Applied analysis of the Navier-Stokes equations.* Cambridge Texts in Applied Mathematics. Cambridge University Press, Cambridge, 1995.

[FP67] C. Foiaş and G. Prodi. Sur le comportement global des solutions non-stationnaires des équations de Navier-Stokes en dimension 2. *Rend. Sem. Mat. Univ. Padova*, 39:1–34, 1967.

[FS84a] C. Foias and J.-C. Saut. Asymptotic behavior, as $t \rightarrow +\infty$, of solutions of Navier-Stokes equations and nonlinear spectral manifolds. *Indiana Univ. Math. J.*, 33(3):459–477, 1984.

[FS84b] C. Foias and J.-C. Saut. On the smoothness of the nonlinear spectral manifolds associated to the Navier-Stokes equations. *Indiana Univ. Math. J.*, 33(6):911–926, 1984.

[FST85] Ciprian Foias, George R. Sell, and Roger Temam. Variétés inertielles des équations différentielles dissipatives. *C. R. Acad. Sci. Paris Sér. I Math.*, 301(5):139–141, 1985.

[FST88] Ciprian Foias, George R. Sell, and Roger Temam. Inertial manifolds for nonlinear evolutionary equations. *J. Differential Equations*, 73(2):309–353, 1988.

[GG05] Isabelle Gallagher and Thierry Gallay. Uniqueness for the two-dimensional Navier-Stokes equation with a measure as initial vorticity. *Math. Ann.*, 332(2):287–327, 2005.

[GGN09] Isabelle Gallagher, Thierry Gallay, and Francis Nier. Special asymptotics for large skew-symmetric perturbations of the harmonic oscillator. *Int. Math. Res. Not. IMRN*, (12):2147–2199, 2009.

[GMO88] Yoshikazu Giga, Tetsuro Miyakawa, and Hirofumi Osada. Two-dimensional Navier-Stokes flow with measures as initial vorticity. *Arch. Rational Mech. Anal.*, 104(3):223–250, 1988.

[GW02] Thierry Gallay and C. Eugene Wayne. Invariant manifolds and the long-time asymptotics of the Navier-Stokes and vorticity equations on $\mathbf{R}^2$. *Arch. Ration. Mech. Anal.*, 163(3):209–258, 2002.

[GW05] Thierry Gallay and C. Eugene Wayne. Global stability of vortex solutions of the two-dimensional Navier-Stokes equation. *Comm. Math. Phys.*, 255(1):97–129, 2005.

[Hal88] Jack K. Hale. *Asymptotic behavior of dissipative systems*, volume 25 of *Mathematical Surveys and Monographs*. American Mathematical Society, Providence, RI, 1988.

[Hen81] Daniel Henry. *Geometric theory of semilinear parabolic equations*, volume 840 of *Lecture Notes in Mathematics*. Springer-Verlag, Berlin, 1981.

[JT93] Don A. Jones and Edriss S. Titi. Upper bounds on the number of determining modes, nodes, and volume elements for the Navier-Stokes equations. *Indiana Univ. Math. J.*, 42(3):875–887, 1993.

[KT01] Y.-J. Kim and A. E. Tzavaras. Diffusive N-waves and metastability in the Burgers equation. *SIAM J. Math. Anal.*, 33(3):607–633 (electronic), 2001.

[KY79] James Kaplan and James Yorke. Chaotic behavior of multidimensional difference equations. In Heinz-Otto Peitgen and Hans-Otto Walther, editors, *Functional Differential Equations and Approximation of Fixed Points*, volume 730 of *Lecture Notes in Mathematics*, pages 204–227. Springer Berlin / Heidelberg, 1979. 10.1007/BFb0064319.

[LL97] Elliott H. Lieb and Michael Loss. *Analysis*, volume 14 of *Graduate Studies in Mathematics*. American Mathematical Society, Providence, RI, 1997.

[Mie91] Alexander Mielke. Locally invariant manifolds for quasilinear parabolic equations. *Rocky Mountain J. Math.*, 21(2):707–714, 1991. Current directions in nonlinear partial differential equations (Provo, UT, 1987).

[Mik98] Milan Miklavčič. *Applied functional analysis and partial differential equations*. World Scientific Publishing Co. Inc., River Edge, NJ, 1998.

[MP76] John Mallet-Paret. Negatively invariant sets of compact maps and an extension of a theorem of Cartwright. *J. Differential Equations*, 22(2):331–348, 1976.

[MPS88] John Mallet-Paret and George R. Sell. Inertial manifolds for reaction diffusion equations in higher space dimensions. *J. Amer. Math. Soc.*, 1(4):805–866, 1988.

[MS01] Tetsuro Miyakawa and Maria Elena Schonbek. On optimal decay rates for weak solutions to the Navier-Stokes equations in $\mathbb{R}^n$. In *Proceedings of Partial Differential Equations and Applications (Olomouc, 1999)*, volume 126, pages 443–455, 2001.

[NSUW09] Raymond Nagem, Guido Sandri, David Uminsky, and C. Eugene Wayne. Generalized Helmholtz-Kirchhoff model for two-dimensional distributed vortex motion. *SIAM J. Appl. Dyn. Syst.*, 8(1):160–179, 2009.

[PP95] A. Prochazka and D. I. Pullin. On the two-dimensional stability of the axisymmetric Burgers vortex. *Phys. Fluids*, 7(7):1788–1790, 1995.

[Rob01] James C. Robinson. *Infinite-dimensional dynamical systems*. Cambridge Texts in Applied Mathematics. Cambridge University Press, Cambridge, 2001. An introduction to dissipative parabolic PDEs and the theory of global attractors.

[TE05] Lloyd N. Trefethen and Mark Embree. *Spectra and pseudospectra*. Princeton University Press, Princeton, NJ, 2005. The behavior of nonnormal matrices and operators.

[UEWB12] David Uminsky, C. Eugene Wayne, and Alethea Barbaro. A multi-moment vortex method for 2D viscous fluids. *J. Comput. Phys.*, 231(4):1705–1727, 2012.

[VI92] A. Vanderbauwhede and G. Iooss. Center manifold theory in infinite dimensions. In *Dynamics reported: expositions in dynamical systems*, volume 1 of *Dynam. Report. Expositions Dynam. Systems (N.S.)*, pages 125–163. Springer, Berlin, 1992.

[Vil09] Cédric Villani. Hypocoercivity. *Mem. Amer. Math. Soc.*, 202(950):iv+141, 2009.

[Way11] C. Eugene Wayne. Vortices and two-dimensional fluid motion. *Notices Amer. Math. Soc.*, 58(1):10–19, 2011.

[YMC03] Z. Yin, D. C. Montgomery, and H. J. H. Clercx. Alternative statistical-mechanical descriptions of decaying two-dimensional turbulence in terms of "patches" and "points". *Phys. Fluids*, 15:1937–1953, 2003.

Chapter 2
Localized States and Dynamics in the Nonlinear Schrödinger/Gross-Pitaevskii Equation

Michael I. Weinstein

1 Introduction

Nonlinear dispersive waves are wave phenomena resulting from the interacting effects of nonlinearity and dispersion. *Dispersion* refers to the property that waves of different wavelengths travel at different velocities. This property may, for example, be due to the material properties of the medium, *e.g.* chromatic dispersion [1], or to the geometric arrangement of material constituents, *e.g.* a periodic medium with Floquet-Bloch *band dispersion* [3]. Nonlinearity distorts the shape of a localized structure or creates amplitude dependent non-uniformities in phase. This may concentrate or localize energy in a region of space (attractive/focusing nonlinearity) or tend to expel energy from compact sets (repulsive/defocusing nonlinearity). In electromagnetics, nonlinearities may arise due to the intensity dependence of dielectric parameters (*e.g.*, Kerr effect [1]). Physical phenomena in which the effects of both dispersion and nonlinearity play a role are ubiquitous. Some examples are: (a) long waves of small amplitude at a water–air interface [114], (b) a nearly mono-chromatic laser beam propagating through air, glass, or water [36], (c) light-pulses propagating through optical fiber waveguides [1], and (d) the macroscopic dynamics of weakly correlated quantum particles in a Bose-Einstein condensate; see, for example, [30, 49, 88]. Interest in nonlinear dispersive waves and their interaction with nonhomogeneous media ranges from Fundamental to Applied Science with

This work is supported in part by U.S. NSF grant DMS-10-08855 and DMS-14-12560.

M.I. Weinstein (✉)
Department of Applied Physics and Applied Mathematics, and Department of Mathematics
Columbia University, New York, NY 10027, USA
e-mail: miw2103@columbia.edu

© Springer International Publishing Switzerland 2015   41
C.E. Wayne, M.I. Weinstein, *Dynamics of Partial Differential Equations*,
Frontiers in Applied Dynamical Systems: Reviews and Tutorials 3,
DOI 10.1007/978-3-319-19935-1_2

great promise in engineering/technological applications due to major advances in materials science and micro- and nano-structure fabrication techniques; see, for example, [109].

An important feature of many nonlinear dispersive systems is the prevalence of coherent structures. They range from phenomena we can only passively observe in nature, such as large-scale localized meteorological events, to those we can control, such as the flow of energy in computer chips and other nano-patterned media. In general, a coherent structure is one which has a spatially localized core, *e.g.* a solitary wave concentrated in a compact region of space or a front-like state which transitions between different asymptotic states across a spatially compact transition region.

We focus on a key model in the mathematical theory of nonlinear dispersive waves, the *nonlinear Schrödinger/Gross Pitaevskii* (NLS/GP) equation:

$$i\partial_t \Phi = -\Delta\Phi + V(x)\Phi + g|\Phi|^2\Phi . \tag{1.1}$$

In particular, we consider its *nonlinear bound states*. Here, $\Phi(x, t)$ is a complex-valued function for $x \in \mathbb{R}^d$ and $t \in \mathbb{R}$. $V(x)$ is a real-valued "potential," and g is a real parameter. The following table lists the physical significance of Φ, V and g in the settings of nonlinear optics and macroscopic quantum physics:

	Nonlinear Optics	Macroscopic Quantum Physics
$\Phi(x, t)$	Electric field complex envelope	Macroscopic quantum wave function
$V(x)$	Refractive index	Magnetic trap
g	Kerr nonlinearity coefficient, $g < 0$	Microscopic 2-body scattering length $(g > 0$ or $g < 0)$

1.1 Outline

Basic mathematical properties of NLS and NLS/GP are discussed in section 2. In section 3 nonlinear bound states and aspects of their stability theory are discussed from variational and bifurcation perspectives. Examples are also presented and the particular cases of $V(x)$ given by a single-well, double-well potential and periodic potential are discussed in detail. In section 4 we discuss particle-like dynamics of solitons interacting with a potential over a large, but finite, time interval. In sections 5 and 6 we consider the very long time behavior of solutions to NLS/GP. In particular, the focus is placed on the important resonant radiation damping mechanism that drives the relaxation of the system to a nonlinear ground state and underlies the phenomena of *Ground State Selection* and *Energy Equipartition*. Sections 5.1–5.2 discuss linear and nonlinear *toy minimal models*, which illustrate these mechanisms. Regarding the overall style of this chapter, we seek to emphasize the key ideas, and therefore do not present all detailed technical hypotheses, leaving that to the references.

2 NLS and NLS/GP

In this section we review basic properties of NLS/GP:

$$i\partial_t \Phi = -\Delta\Phi + V(x)\Phi + g|\Phi|^{2\sigma}\Phi. \tag{2.1}$$

Here, g is a real parameter and we assume that V is a bounded and smooth real-valued potential. The equation (2.1) is often called the nonlinear Schrödinger/Gross-Pitaevskii (NLS/GP) equation.

We consider solutions to the initial value problem for (2.1) with initial conditions

$$\Phi(x, 0) = \Phi_0(x) \in H^1(\mathbb{R}^d). \tag{2.2}$$

The basic well-posedness theorem states that for $d = 1, 2$ and all $0 \le \sigma < \infty$ and for $d \ge 3$ and $0 \le \sigma < 2/(d-2)$, the initial value problem (2.1)–(2.2) has a unique local-in-time solution $\Phi \in C^0([0, T); H^1(\mathbb{R}^d))$ [6, 37, 64, 102, 103].

NLS/GP is a Hamiltonian system, expressible in the form

$$i\partial_t \Phi = \frac{\delta\mathcal{H}[\Phi, \Phi^*]}{\delta\Phi^*}, \tag{2.3}$$

with Hamiltonian energy functional

$$\mathcal{H}[F, F^*] = \int_{\mathbb{R}^d} \nabla F(x) \cdot \nabla F^*(x) + V(x)\, F(x)F^*(x) + \frac{g}{\sigma+1}\left(F(x)F^*(x)\right)^{\sigma+1}\, dx. \tag{2.4}$$

The solution of the initial value problem satisfies the conservation laws:

$$\mathcal{H}[\Phi(\cdot, t)] = \mathcal{H}[\Phi_0] \tag{2.5}$$

$$\mathcal{N}[\Phi(\cdot, t)] = \mathcal{N}[\Phi_0] \tag{2.6}$$

where $\mathcal{H}[\cdot]$ is defined in (2.4) and

$$\mathcal{N}[F, F^*] = \int_{\mathbb{R}^d} F(x)F^*(x)\, dx. \tag{2.7}$$

The conserved integrals $\mathcal{H}$ and $\mathcal{N}$ are associated, respectively, with the invariances $t \mapsto t + t_0$, $t_0 \in \mathbb{R}$ (time-translation) and $\Phi \mapsto \Phi e^{i\theta}$, $\theta \in \mathbb{R}$.

For the nonlinear Schrödinger equation (NLS), where $V \equiv 0$, we have the additional symmetries

1. $\Phi(x, t) \mapsto \Phi(x + x_0, t)$, $x_0 \in \mathbb{R}^d$ (Translation invariance)
2. $\Phi(x, t) \mapsto \mathcal{G}_\xi[\Phi](x, t) = e^{i\xi\cdot(x-\xi t)}\Phi(x - 2\xi t, t)$, $\xi \in \mathbb{R}^d$ (Galilean invariance)

For NLS with the homogeneous power nonlinearity in (2.1) the equation is also dilation invariant:

$$\Phi(x,t) \mapsto \lambda^{\frac{1}{\sigma}} \Phi(\lambda x, \lambda^2 t) \tag{2.8}$$

The case where $g < 0$ is called the *focusing case* and the case where $g > 0$ is called the *defocusing case*. This terminology refers to the nonlinear term's tendency to focus (localize) energy or defocus (spread) energy. Independently of the nonlinear term, the linear potential may be an attractive potential, which concentrates energy, or a repulsive potential which allows energy to spread.

Concerning global-in-time behavior of solutions, the initial value problem is globally well-posed for any H^1 initial condition if either $g > 0$ or if $g < 0$ and $\sigma < 2/d$ (subcritical case). For $g < 0$ and $\sigma \geq 2/d$ solutions with H^1 initial data may develop singularities in finite time (blow up). See, for example, [6, 37, 64, 102, 103, 110].

3   Bound States - Linear and Nonlinear

3.1   Linear bound states

We start with the linear Schrödinger equation:

$$i\partial_t \Psi = (-\Delta + V(x)) \Psi = H \Psi \tag{3.1}$$

Here V denotes a smooth and real-valued potential satisfying

$$V(x) \to 0 \text{ sufficiently rapidly as } |x| \to \infty . \tag{3.2}$$

Bound states are solutions of the form $\Psi = e^{-iEt} \psi(x)$, where ψ satisfies the eigenvalue problem

$$H\psi = (-\Delta + V)\psi = E\psi, \ \psi \in H^1(\mathbb{R}^d) . \tag{3.3}$$

We expect the eigenvalue problem (3.3) to have a nontrivial solutions if V is an attractive "potential well"; intuitively, the set where $V(x) \leq 0$ should be sufficiently wide and deep.

The following result gives a condition on the existence of a ground state, a nontrivial solution of (3.3) of minimal energy, E [15, 78].

Theorem 3.1. *Denote by*

$$E_{0\star} = \inf_{\int_{\mathbb{R}^d} |f|^2 = 1} \langle Hf, f \rangle \equiv \inf_{\int_{\mathbb{R}^d} |f|^2 = 1} \int_{\mathbb{R}^d} |\nabla f(x)|^2 + V(x)|f(x)|^2 \, dx.$$

If $-\infty < E_{0\star} < 0$, *then the infimum is attained at a positive function* $\psi_{0\star} \in H^1(\mathbb{R}^d)$, *which is a solution of*

$$H\psi_{0\star} = E_{0\star}\psi_{0\star}, \quad E_{0\star} = \langle H\psi_{0\star}, \psi_{0\star}\rangle$$

3.2 Nonlinear bound states

Here we consider spatially localized solutions of NLS/GP of the form

$$\Phi(x,t) = e^{-iEt}\psi_E(x), \tag{3.4}$$

where ψ_E satisfies the nonlinear elliptic equation

$$(-\Delta + V)\psi_E + g|\psi_E|^{2\sigma}\psi_E = E\psi_E, \quad \psi_E \in H^1(\mathbb{R}^d) \tag{3.5}$$

Solutions of the nonlinear eigenvalue problem (3.5) are called *nonlinear bound states* or solitary waves. A positive and decaying solution is called a nonlinear ground state. In the case where $V(x)$ is nontrivial, such states are often called nonlinear defect modes. In this case, the spatially varying potential, $V(x)$, is viewed as a localized defect in a homogeneous (translation-invariant) background.

The question of existence of nonlinear bound states of NLS/GP has been studied extensively by variational and bifurcation methods. See, for example, [5, 79, 90, 101, 110] as well as the discussion in sections 3.6 [73] and 3.7 [59].

The following result, which we shall refer to later in this chapter, concerns bifurcation of a nonlinear ground state from the ground eigenstate of the linear operator $-\Delta + V$; see, for example, [87, 90]

Theorem 3.2. *Assume V is real-valued, smooth, and rapidly decreasing at infinity and that* $-\Delta + V$ *has at least one discrete eigenvalue. Let* $(\psi_{0\star}, E_{0\star})$ *denote the ground state eigenpair given by Theorem 3.1; here,* $\psi_{0\star}$ *may be chosen to be positive and normalized in* L^2. *Consider the equation for nonlinear bound states of NLS/GP with* $g = -1$ *(attractive nonlinear potential):*

$$\left(-\Delta + V(x) - |\psi_E|^{2\sigma}\right)\psi_E = E\psi_E, \quad E < 0, \quad \sigma \geq 1. \tag{3.6}$$

Then, there is a constant $\delta_0 > 0$ *and a non-empty interval* $\mathcal{I} = (E_{0\star} - \delta_0, E_{0\star})$ *such that for any* $E \in \mathcal{I}$, *equation* (3.6) *has a positive solution, which for E tending to* $E_{0\star}$ *has the expansion:*

$$\psi_E(x) = \rho(E)\left(\psi_{0\star}(x) + \mathcal{O}\left(\rho(E)^{2\sigma}\right)\right), \quad \rho(E) = |E_{0\star} - E|^{\frac{1}{2}}\left(\int \psi_{0\star}^{2\sigma+2}\right)^{-\frac{1}{2\sigma}} \tag{3.7}$$

3.3   Orbital stability of nonlinear bound states

In this section we discuss the *orbital Lyapunov stability* of nonlinear grounds states. In particular, we prove nonlinear stability in $H^1(\mathbb{R}^n)$ modulo the natural group of symmetries. We shall in this section consider the case where $g = -1$, the case of focusing nonlinearity. Given a positive H^1 solution of (3.6) introduce its *orbit*:

$$\mathcal{O}_{gs} = \left\{\Psi_E(x)e^{i\gamma} : \gamma \in [0, 2\pi)\right\}, \quad V \neq 0$$

$$\mathcal{O}_{gs} = \left\{\Psi_E(x + x_0)e^{i\gamma} : \gamma \in [0, 2\pi), \ x_0 \in \mathbb{R}^n\right\}, \quad V \equiv 0. \tag{3.8}$$

We say that the ground state is *orbitally stable* in $H^1(\mathbb{R}^n)$ if the following holds. If $\Phi(x, t = 0)$ is close to element of $\mathcal{O}_{gs}$ in H^1, then for all $t \neq 0$, $\Phi(x, t)$ is H^1 close to some (typically t dependent) element of $\mathcal{O}_{gs}$. In order to make this precise, we introduce a metric which measures the distance from an arbitrary H^1 function to the ground state orbit:

$$\text{dist}\left(u, \mathcal{O}_{gs}\right) = \inf_{\gamma} \|u - \Psi_0 e^{i\gamma}\|_{H^1}, \quad V \neq 0$$

$$\text{dist}\left(u, \mathcal{O}_{gs}\right) = \inf_{\gamma, x_0} \|u(\cdot + x_0) - \Psi_0 e^{i\gamma}\|_{H^1}, \quad V \equiv 0. \tag{3.9}$$

A more precise statement of H^1- Lyapunov stability is as follows. For any $\epsilon > 0$ there is a $\delta > 0$ such that if

$$\text{dist}\left(\Phi(\cdot, 0), \mathcal{O}_{gs}\right) < \delta \tag{3.10}$$

then for all $t \neq 0$

$$\text{dist}\left(\Phi(\cdot, t), \mathcal{O}_{gs}\right) < \epsilon.$$

Nonlinear bound states are critical points of the Hamiltonian $\mathcal{H}$ subject to fixed L^2 norm, $\mathcal{N}$. In particular, a nonlinear bound state, ψ_E, of frequency E satisfies $\delta \mathcal{E}_E[f, f^*]/\delta f^* = 0$, where

$$\mathcal{E}_E[f] = \mathcal{H}[f] - E \int |f|^2$$

$$\mathcal{H}[f] \equiv \int |\nabla f|^2 + V(x)|f|^2 - \frac{1}{\sigma + 1}|f|^{2\sigma+2}$$

For subcritical nonlinearities, $\sigma < 2/d$, stable nonlinear ground states may be realized as constrained global minimizers of the variational problem:

$$\inf_{f \in H^1} \mathcal{H}[f] \qquad \text{subject to fixed } \mathcal{N}[f] > 0. \tag{3.11}$$

Orbital H^1 stability is a consequence of the compactness properties of arbitrary minimizing sequences [12, 79].

However in general, depending on the details of the potential and nonlinear terms, stable solitary waves may arise as *local* minimizers of $\mathcal{H}$ subject to fixed $\mathcal{N}$. We now discuss stability, in this more general setting. Introduce the linearized operators $L_\pm$, which are real and imaginary parts of the second variation of the energy $\mathcal{E}_E$:

$$L_+ \equiv -\Delta + V(x) - (2\sigma + 1)\psi_E^{2\sigma} - E. \tag{3.12}$$

$$L_- \equiv -\Delta + V(x) - \psi_E^{2\sigma} - E \tag{3.13}$$

We define the *index of ψ_E* by

$$\text{index}\,(\psi_E) \; = \text{number of strictly negative eigenvalues of } L_+ \,. \tag{3.14}$$

Theorem 3.3 (H^1 **Orbital Stability**).

1. Assume the conditions

(S1) Spectral condition: $\psi_E > 0$ and index $(\psi_E) = 1$
(S2) Slope condition

$$\frac{d}{dE}\mathcal{N}[\psi_E] \; \equiv \; \frac{d}{dE}\int |\psi_E(x)|^2 dx \; < \; 0 \tag{3.15}$$

Then, ψ_E is H^1 orbitally stable.

2. Assume $\psi_E > 0$. If index $(\psi_E) \geq 2$ *or $\frac{d}{dE}\mathcal{N}[\psi_E] > 0$, then ψ_E is linearly exponentially unstable. That is, the linearized evolution equation (see (3.19) below) has a spatially localized solution which grows exponentially with t.*

See [50, 51, 62, 93, 111, 112]. In [108] it was shown that $\partial_E \mathcal{N}[\psi_E] > 0$ implies the existence of an exponentially growing mode of the linearized evolution equation.

We give the idea of the proof of stability. For simplicity, suppose $V(x)$ is nontrivial, In this case the ground state orbit consists of all phase translates of ψ_E; see the first line of (3.8). Let ϵ be an arbitrary positive number. We have for $t \neq 0$, by choosing δ in (3.10) sufficiently small

$$\epsilon^2 \; \sim \; \mathcal{E}_E[\Psi(\cdot,0)] \; - \; \mathcal{E}_E[\psi_E]$$

$$= \; \mathcal{E}_E[\Psi(\cdot,t)] \; - \; \mathcal{E}_E[\psi_E], \quad \text{by conservation laws,}$$

$$= \; \mathcal{E}_E[\Psi(\cdot,t)e^{i\gamma}] \; - \; \mathcal{E}_E[\psi_E], \quad \text{by phase invariance}$$

$$= \; \mathcal{E}_E[\psi_E + u(\cdot,t) + iv(\cdot,t)] \; - \; \mathcal{E}_E[\psi_E]$$

(definition of the perturbation $u + iv, u, v \in \mathbb{R}$)

$$\sim \; \langle L_+ u(t), u(t)\rangle + \langle L_- v(t), v(t)\rangle \quad \text{(by Taylor expansion and } \delta\mathcal{E}_E[\psi_E]=0) \tag{3.16}$$

If L_+ and L_- were positive definite operators, implying the existence of positive constants C_+ and C_- such that

$$\langle L_+ u, u \rangle \geq C_+ \|u\|_{H^1}^2, \tag{3.17}$$

$$\langle L_- v, v \rangle \geq C_- \|v\|_{H^1}^2 \tag{3.18}$$

for all $u, v \in H^1$, then it would follow from (3.16) that the perturbation about the ground state, $u(x,t) + iv(x,t)$, would remain of order ϵ in H^1 for all time $t \neq 0$. The situation is however considerably more complicated. The relevant facts to note are as follows.

(1) $L_- \psi_E = 0$, with $\psi_E > 0$. Hence, ψ_E is the ground state of L_-, $0 \in \sigma(L_-)$, and L_- is a non-negative with continuous spectrum $[\,|E|, \infty)$.
(2) For small L^2 nonlinear ground states, L_+ has exactly one strictly negative eigenvalue and continuous spectrum $[\,|E|, \infty)$.

The zero eigenvalue of L_- and the negative eigenvalue of L_+ constitute two *bad* directions, which are treated as follows, noting that $u(\cdot, t)$ and $v(\cdot, t)$ are not arbitrary H^1 functions but are rather constrained by the dynamics of NLS.

To control L_-, we choose $\gamma(t)$ so as to minimize the distance of the solution to the ground state orbit, (3.9). This yields the codimension one constraint on v: $\langle v(\cdot, t), \psi_E \rangle = 0$, subject to which (3.18) holds with $C_- > 0$.

To control L_+, we observe that since the L^2 norm is invariant for solutions, we have the codimension constraint on u: $\langle u(\cdot, t), \psi_E \rangle = 0$. Although ψ_E is not the ground state of L_+, it can be shown by constrained variational analysis that if the slope condition (3.15) holds, then constraint on u places u in the positive cone of L_+, i.e. $C_+ > 0$ in (3.17). Thus, positivity (coercivity) estimates (3.17) and (3.18) hold and $\mathcal{E}_E$ serves as a Lyapunov functional which controls the distance of the solution to the ground state orbit. The detailed argument is presented in [112].

We also note that the role played by the second variation of $\mathcal{E}$ in the linearized time-dynamics [111]. Let $\Psi(t) = (\psi_E + u + iv)e^{-iEt}$. The linearized perturbation, $(u(t), v(t))^t$, satisfies the linear Hamiltonian system:

$$\partial_t \begin{pmatrix} u \\ v \end{pmatrix} = \begin{pmatrix} 0 & L_- \\ -L_+ & 0 \end{pmatrix} \begin{pmatrix} u \\ v \end{pmatrix}, \tag{3.19}$$

with conserved (time-invariant) energy

$$\mathcal{Q}(u, v) \equiv \langle L_+ u, u \rangle + \langle L_- v, v \rangle \tag{3.20}$$

The above constrained variational analysis underlying nonlinear orbital theory corresponds to the H^1 boundedness of the linearized flow (3.19), restricted on a finite codimension subspace. This subspace is expressible in terms of symplectic orthogonality to first order generators of symmetries acting on the ground state.

This structure is used centrally in many works on nonlinear asymptotic stability and nonlinear scattering theory of solitary waves; see section 6 and the references cited.

In the following subsections, we give several examples: a) the free NLS soliton, b) the nonlinear defect mode of a simple potential well, c) nonlinear bound states and symmetry breaking for the double well, and d) *gap solitons* of NLS/GP with a periodic potential.

3.4 The free soliton of focusing NLS: $V \equiv 0$ and $g = -1$

In this case, there is a unique (up to spatial translation) positive and symmetric solution, a ground state, which is smooth and decays exponentially as $|x| \to \infty$. For results on existence, symmetry, and uniqueness of the ground state, see e.g. [5, 44, 76, 101].

For example, the focusing one-dimensional cubic ($\sigma = 1$) nonlinear Schrödinger equation

$$i\partial_t \psi = -\partial_x^2 \psi - |\psi|^2 \psi, \qquad (3.21)$$

has the solitary standing wave:

$$\psi_{sol}(x, t) = e^{it} \sqrt{2} \operatorname{sech}(x) , \qquad (3.22)$$

By the symmetries of NLS (section 2), we have the extended family of solitary traveling waves of arbitrary negative frequency $E = -\lambda^2$, $\lambda > 0$:

$$\mathcal{G}_{\lambda, x_0, v, \theta} [\psi_{sol}](x, t) = \lambda e^{i\lambda^2 t} \sqrt{2} \operatorname{sech}(\lambda (x - x_0 - 2vt)) \, e^{iv(x - x_0 - vt)} \, e^{i\theta}; \quad v, \theta, x_0 \in \mathbb{R}. \qquad (3.23)$$

Theorem 3.3 implies that the positive solitary standing wave of translation invariant NLS is orbitally stable if $\sigma < 2/d$ and is unstable if $\sigma \geq 2/d$ [51, 110, 112]. The solid curve in Figure 2.1 shows the family of solitary waves, (3.23), of (3.21) bifurcating from the zero solution at the edge-energy of the continuous spectrum of $-\partial_x^2$.

3.5 $V(x)$, a simple potential well; model of a pinned nonlinear defect mode

In [90] the bifurcation of nonlinear bound states of NLS/GP from localized eigenstates of the linear Schrödinger operator $-\Delta + V$ was studied. The simplest

Figure 2.1 Squared L^2
norm, $\mathcal{N}$, vs. frequency for
the free soliton (solid) and
pinned soliton (dashed).

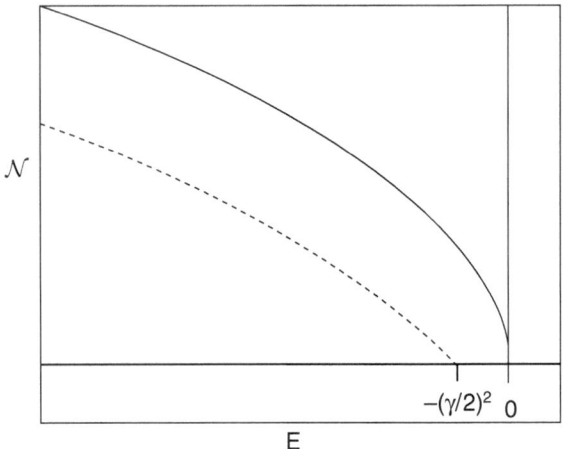

case of such a *nonlinear defect mode* is for NLS/GP with $V(x)$ is taken to be a Dirac
delta function potential well:

$$i\partial_t\psi \;=\; H\psi - |\psi|^2\psi, \;\; H = -\partial_x^2 \,-\, \gamma\delta(x), \;\; \gamma > 0 \,. \tag{3.24}$$

For the well-posedness theory of the initial-value-problem for (3.24) in $C^0([0, T);$
$H^1(\mathbb{R}))$ see [60] and, more specifically, section 8D of [27]. In this case, the nonlinear
defect mode of frequency $E = -\lambda^2 < -(\gamma/2)^2$ is explicitly given by:

$$\psi_{sol}(x, t; \gamma) \;=\; \lambda e^{i\lambda^2 t}\sqrt{2}\,\mathrm{sech}\left(\lambda|x| + \tanh^{-1}\frac{\gamma}{\lambda}\right)\,e^{i\theta}$$

This family of nonlinear bound states, which is pinned to the "defect" at $x = 0$,
bifurcates from the zero solution at $E = E_\star = -(\gamma/2)^2$, the unique negative
eigenvalue of H; see the dashed curve in Figure 2.1.

3.6   NLS/GP: Double-well potential with separation, L

Consider now the NLS/GP with a double-well potential, $V_L(x)$:

$$i\partial_t\psi \;=\; -\Delta\psi \,-\, V_L(x)\psi - |\psi|^2\psi \,.$$

We take the double-well, $V_L(x)$ to be constructed by centering two identical single-
bound state potentials, $V_0(x)$, of the sort considered in the example of section 3.5, at
the positions $x = \pm L$:

$$V_L(x) \;=\; V_0(x + L) + V_0(x - L) \tag{3.25}$$

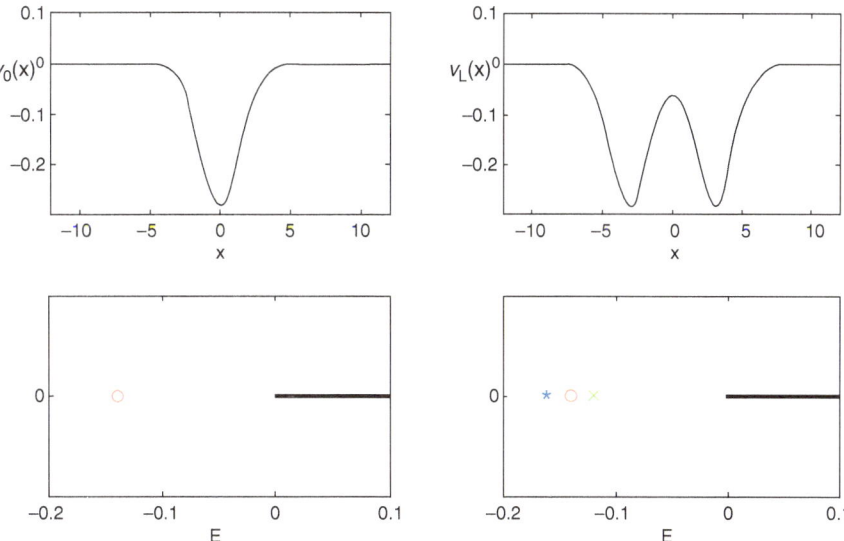

Figure 2.2 Left: Single-well potential and its spectrum: one discrete eigenvalue, marked by "o", and continuous spectrum, $\mathbb{R}_+$. Right: Double-well potential and its spectrum: two nearby discrete eigenvalues (marked "*" and "×") and continuous spectrum, $\mathbb{R}_+$

A sketch of a one-dimensional double-well potential, $V_L(x)$, and the associated spectrum of $H = -\partial_x^2 + V_L(x)$ is displayed in the right panels of Figure 2.2.

As in the example of section 3.5 (see [90]) there are branches of nonlinear bound states which bifurcate from the zero solution at the discrete eigenvalue energies. We focus on the branch emanating from the zero solution at the ground state eigenvalue of H_L. For small squared L^2 norm, $\mathcal{N}$, this is the unique (up to the symmetry $\psi \mapsto e^{i\theta}\psi$) nontrivial solution branch. This solution has the same symmetries as the ground state of the linear double-well potential [52]. That is, for $\mathcal{N}$ small, $\psi_{E(\mathcal{N})}$ is bi-modal with peaks centered at $x \pm L$. Increasing the L^2 norm (or its square, $\mathcal{N}$) we find that there is a critical value $\mathcal{N}_{cr} > 0$, such that for $\mathcal{N} > \mathcal{N}_{cr}$ there are multiple nonlinear bound states; see Figure 2.3. In particular, at $(E(\mathcal{N}_{cr}), \psi_{\mathcal{N}_{cr}})$ there is a symmetry breaking bifurcation. Specifically, for $\mathcal{N} > \mathcal{N}_{cr}$, there are three families of solutions: the continuation of the symmetric branch (dashed curve continuation of the symmetric branch) and two branches of asymmetric states, corresponding to solutions whose mass is concentrated on the left or right side wells. The bifurcation diagram in Figure 2.3 shows only two branches beyond the bifurcation point. The solid leftward branching curve represents both asymmetric branches, one set of states being obtained from the other via a reflection about $x = 0$.

Figure 2.3 also encodes stability properties. For $0 < \mathcal{N} < \mathcal{N}_{cr}$, the (symmetric) ground state is stable, while for $\mathcal{N} > \mathcal{N}_{cr}$ the symmetric state is unstable. At $\mathcal{N} = \mathcal{N}_{cr}$ stability is transferred to the asymmetric branches (solid curve for $\mathcal{N} > \mathcal{N}_{cr}$).

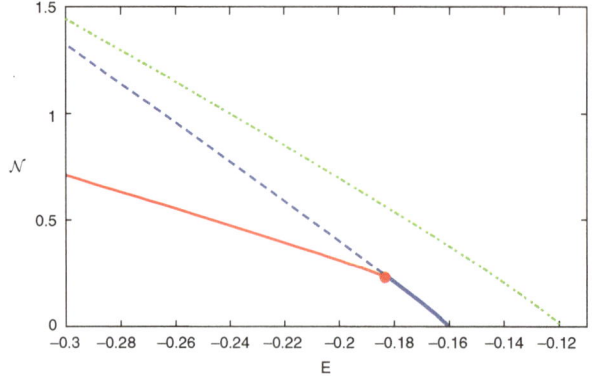

Figure 2.3 Bifurcations from the two discrete eigenvalues of the double-well: Bifurcation curve emanating from the zero state at the ground state (lowest) energy shows symmetry breaking at some positive $\mathcal{N}_{cr}$. Bifurcation curve emanating from the zero state at the excited state energy shows no symmetry breaking.

The above discussion is summarized in the following [66, 73]:

Theorem 3.4 (Symmetry Breaking Bifurcation). *Consider the nonlinear eigenvalue problem*

$$(-\Delta + V_L)\psi_E + g|\psi_E|^2\psi_E = E\psi_E, \quad \psi_E \in H^1(\mathbb{R}^d) \tag{3.26}$$

where $g < 0$ and V_L denotes a double-well potential with separation parameter, L, as in (3.25). Denote the ground and excited state eigenvalues of $-\Delta + V_L$ by:

$$E_{0\star}(L) < E_{1\star}(L) < 0. \tag{3.27}$$

Let $\mathcal{N}[f] = \int_{\mathbb{R}^d} |f|^2$.
   There exists a positive constant, L_0, such that the following holds. For all $L \geq L_0$, there exists $\mathcal{N}_{cr} = \mathcal{N}_{cr}(L) > 0$ such that

(a) For any $\mathcal{N} < \mathcal{N}_{cr}$ there is a unique nontrivial symmetric state.
(b) The point $(E, \psi_E) = (E_{cr}, \psi_{E_{cr}})$ is a bifurcation point, i.e.
   there are, for $\mathcal{N} > \mathcal{N}_{cr}$, two bifurcating branches asymmetric states, consisting, respectively, of states concentrated about $x = \pm L$.
(c) Concerning the stability of these branches:

   (c1) $\mathcal{N} < \mathcal{N}_{cr}$: The symmetric branch is orbitally stable.
   (c2) $\mathcal{N} > \mathcal{N}_{cr}$: The symmetric branch is linearly exponentially unstable; these states have index $= 2$; see (3.14).
   (c3) $\mathcal{N} > \mathcal{N}_{cr}$: The asymmetric branch is orbitally stable.

(d) Symmetry breaking threshold:

$$\mathcal{N}_{cr}(L) \sim E_{1\star}(L) - E_{0\star}(L) \tag{3.28}$$

(The gap, $E_{1\star}(L) - E_{0\star}(L)$ is exponentially small for large L [52, 60].)

Theorem 3.4 is proved by methods of bifurcation theory [45, 84]. Specifically, for L sufficiently large we can use a Lyapunov-Schmidt reduction strategy to reduce the nonlinear eigenvalue problem to a weakly perturbed system of two nonlinear algebraic equations depending on $\mathcal{N}$. The unknowns of this system are essentially the projections of the sought nonlinear bound state onto the linear ground and excited states of H_L. The bifurcation structure of this pair of nonlinear algebraic equations is of the type displayed in Figure 2.3. This bifurcation structure is then shown to persist under the (infinite dimensional) perturbing terms to this finite algebraic reduction. The latter are controlled via PDE estimates and an implicit function theorem argument.

An illustrative study of the *global bifurcation structure* for the exactly solvable case where $V_L(x)$ is taken to be a sum of two attractive Dirac-delta wells centered at $x = \pm L$ is presented in [60]. In [63] the local bifurcation methods described above are used to study the NLS/GP with a triple well potential. A recent study of global properties of the bifurcating branches and their stability properties for general nonlinearities, $|\psi|^{2\sigma}\psi$, with $\sigma \geq 1/2$ is presented in [66]. In this work the authors prove, for the case of symmetric double-well potentials, that as L varies a symmetry breaking bifurcation occurs once the potential develops a local maximum at the origin. The analysis in [66] allows for the bifurcation to occur at large $\mathcal{N}_{cr}$, outside the weakly nonlinear regime where the analysis in [73] applies.

Local bifurcation methods can be applied as well to study excited state branches. An example is the branch emanating from the linear excited state energy, along which it can be shown, for small $\mathcal{N}$, that there are no secondary bifurcations; see Figure 2.3 and the corresponding mode curve which changes sign in Figure 2.4.

An approach which complements the local bifurcation approach is variational. For example, in [2] symmetry breaking in a three-dimensional nonlinear Hartree

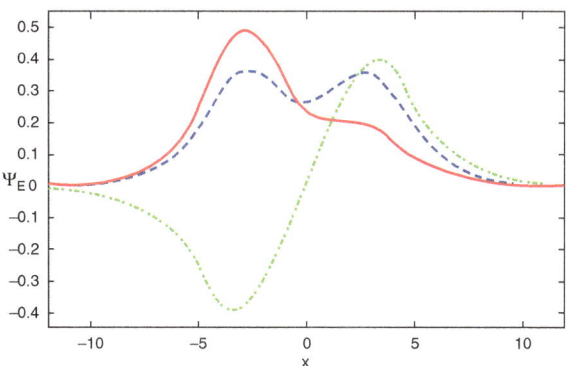

Figure 2.4 Symmetric, asymmetric, and anti-symmetric modes arising in the bifurcation diagram of Figure 2.3.

model having an attractive non-local potential is proved for $\mathcal{N}$ sufficiently large. Their arguments are straightforwardly adaptable to the variational formulation (3.11) and yield that for $\sigma < 2/d$ the ground state is asymmetric. The proof is based on a trial function argument and the intuition that for $\mathcal{N}$ sufficiently large it is energetically preferable to concentrate mass in the left well or in the right well but not both.

3.7 NLS/GP: $V(x)$ periodic and the bifurcations from the spectral band edge

We now consider the nonlinear bounds states of NLS/GP for the case where V is periodic on $\mathbb{R}^d$. In this case, the spectrum of $H = -\Delta + V(x)$ is absolutely continuous. Let ψ_μ denote a nonlinear bound state of frequency μ. Any bifurcation of nontrivial solutions of (3.5) from the zero solution must occur at an energy in the continuous spectrum. Indeed, the example of the free soliton of section 3.4 is an example; $V \equiv 0$ is periodic (!) and has continuous spectrum equal to the half-line $\mathbb{R}_+ = [0, \infty)$. The soliton bifurcates from the zero-amplitude solution (as measured say in L^∞ or L^2) from the continuous spectral edge (solid curve in Figure 2.3). Indeed, for NLS with $V \equiv 0$ and power nonlinearity $-|\psi|^{2\sigma}\psi$

$$\psi_\mu(x) = (-\mu)^{\frac{1}{2\sigma}} \psi_{-1}(\sqrt{-\mu}x) \tag{3.29}$$

and therefore

$$\mathcal{N}[\psi_\mu] = \|\psi_\mu\|_{L^2}^2 = (-\mu)^{\frac{1}{\sigma}-\frac{d}{2}} \|\psi_{-1}\|_{L^2} . \tag{3.30}$$

For $V(x)$ a general periodic potential the spectrum of $H = -\Delta + V$ is the union of spectral bands obtained as follows. Consider the $d-$ parameter family of periodic elliptic eigenvalue problems:

$$\left(-(\nabla + ik)^2 + V(x) \right) u(x) = \mu \, u(x)$$

$$u(x) \text{ is periodic with the periodicity of } V(x) , \tag{3.31}$$

where k varies over a fundamental dual lattice cell, the Brillouin zone, $\mathcal{B} \subset \mathbb{R}^d$. For each fixed $k \in \mathcal{B}$, the spectrum of (3.31) is discrete and consists of eigenvalues denoted:

$$\mu_1(k) \leq \mu_2(k) \leq \cdots \leq \mu_b(k) \leq \cdots , \tag{3.32}$$

listed with multiplicities and tending to positive infinity. The corresponding eigen-functions are denoted, $p_j(x; k)$. The states $\{e^{ikx}p_j(x; k)\}$ where $j \geq 1$ and $k \in \mathcal{B}$ are complete in $L^2(\mathbb{R}^d)$.

It can be shown that nonlinear bound states bifurcate from edges of the spectral bands: to the left, if the nonlinearity is attractive ($g < 0$) and to the right if the nonlinearity is repulsive ($g > 0$) [59, 86, 91]. We expect similar scaling of the L^2 norm of such more general edge-bifurcations:

$$\mathcal{N}[\psi_\mu] = \|\psi_\mu\|^2_{L^2} \sim |\mu - \mu_\star|^{\frac{1}{\sigma} - \frac{d}{2}} \tag{3.33}$$

where $|\mu - \mu_\star|$ is the distance to the spectral band edge located at $\mu_\star$. Now the details of the bifurcation at the edge depend on the periodic structure. We first give a heuristic picture and then state a precise theorem.

For $|\mu - \mu_\star|$ small, that is for μ near a spectral band edge, the nonlin-ear bound state, $\psi_\mu(x)$, should be exponentially localized with decay $\psi_\mu(x) \sim \exp(-|\mu - \mu_\star|^{\frac{1}{2}}|x|)$ for $|x| \to \infty$. Due to the separation of length scales: $|\mu - \mu_\star|^{-1} \gg$ period of V, we expect $\psi_\mu(x)$ should oscillate like the Floquet-Bloch mode $p_1(x; 0) = p_\star(x)$, associated with the band edge energy, $\mu_\star$, with a slowly varying and spatially localized amplitude, $F(x)$, :

$$\psi \sim \delta^{\frac{1}{\sigma}} F(\delta x) \times p_\star(x), \quad \delta = |\mu - \mu_\star|^{\frac{1}{2}}, \tag{3.34}$$

where $F(x)$ satisfies an effective medium (homogenized/constant coefficient) NLS equation.

To simplify the discussion we will focus on bifurcation from the lowest energy (bottom) of the continuous spectrum, $\mu_\star = \mu_1(0)$, for the case of an attractive nonlinearity ($g < 0$). Before stating a precise result, we introduce the inverse effective mass matrix associated with the bottom of the continuous spectrum defined by

$$A_{\text{eff}} \equiv \frac{1}{2}D^2\mu_1(k = 0) = \frac{1}{2}\left(\frac{\partial^2\mu_1(k = 0)}{\partial k_i \partial k_j}\right)_{1 \leq i,j \leq d}$$

$$= \delta_{ij} - \frac{4\left\langle\partial_{x_j}p_\star, (-\Delta + V - \mu_\star)^{-1}\partial_{x_i}p_\star\right\rangle}{\langle p_\star, p_\star\rangle} \tag{3.35}$$

and Γ_{eff}, the effective nonlinearity coefficient defined by

$$\Gamma_{\text{eff}} = \frac{\int p_\star(x)^{2\sigma+2}\,dx}{\int p_\star(x)^2\,dx} > 0. \tag{3.36}$$

The matrix A_{eff} is clearly symmetric and, for the lowest band edge, it is positive definite [74].

The following result, proved in [59], describes the bifurcation from the bottom of continuous spectrum (left end point of the first spectral band). See also [86, 91, 100]. A proof of the case of a general band edge is discussed in [59].

Theorem 3.5 (Edge bifurcations of nonlinear bound states for periodic potentials). *Let $V(x)$ denote a smooth and even periodic potential in dimension $d = 1, 2$ or $d = 3$. Let x_0 denote any local minimum or maximum of $V(x)$. Denote by $\mu_\star = \inf \sigma(-\Delta + V)$. Let $F_{A,\Gamma}(y)$ the unique, centered at $y = 0$, positive $H^1(\mathbb{R}^d)$ solution (ground state) of the effective/homogenized nonlinear Schrödinger equation with inverse effective mass matrix, $A = A_{\mathrm{eff}}$, and effective nonlinearity, $\Gamma = \Gamma_{\mathrm{eff}}$:*

$$-\sum_{i,j=1}^{d} \frac{\partial}{\partial y_i} A_{\mathrm{eff}.ij} \frac{\partial}{\partial y_j} F(y) - \Gamma_{\mathrm{eff}} F^{2\sigma+1}(y) = -F(y) . \tag{3.37}$$

Then, there exists $\delta_0 > 0$ such that for all $\mu \in (\mu_\star - \delta_0, \mu_\star)$ there is a family of nonlinear bound states $\mu \mapsto \psi_\mu(x)$

$$\psi_\mu(x) - (\mu_* - \mu)^{\frac{1}{2\sigma}} F_{A_{\mathrm{eff}},\Gamma_{\mathrm{eff}}} \left(\sqrt{\mu_* - \mu} \, (x - x_0) \right) p_\star(x) \to 0, \quad \text{as } \mu \uparrow \mu_\star \text{ in } H^1(\mathbb{R}^d) . \tag{3.38}$$

The proof of Theorem 3.5 [59], like the study of bifurcations from discrete eigenvalues, proceeds via Lyapunov-Schmidt reduction and application of the implicit function theorem. However, while the analysis of bifurcation from discrete spectrum leads to a finite dimensional (nonlinear algebraic) bifurcation equation, bifurcation from the continuous spectrum leads to an infinite dimensional bifurcation equation, a nonlinear homogenized partial differential equation (3.37); see also [25, 26]. Examples of recent applications of these and related ideas to other systems appear in [28, 29, 31–35, 61].

Since (3.37) is a constant coefficient equation, $F_{A,\Gamma}$ can be related, via scaling, to the nonlinear ground state of the translation invariant nonlinear Schrödinger equation. Thus, the shape of the bifurcation diagram $\mu \mapsto \mathcal{N}[\psi_\mu]$ (see Figure 2.3) can be deduced for μ near $\mu_\star$; see [59]. We also note that the stability/instability properties of ground states of (3.37) for the time-dependent effective nonlinear Schrödinger equation are a consequence of Theorem 3.3 in the translation invariant case (Figure 2.5).

4   Soliton/Defect Interactions

In this section we turn to the detailed time dynamics of a soliton-like nonlinear bound state interacting with a potential. This question has fundamental and applications interest. From the fundamental perspective it is an important problem in the direction of developing a nonlinear scattering theory for non-integrable Hamiltonian PDEs. And, from an applications perspective, nonlinear waves often arise in systems

Figure 2.5 $P = \mathcal{N}[\psi_\mu]$ vs. μ for the 1d - NLS/GP with periodic potential, $V(x)$ for the case $\sigma = \sigma_{\mathrm{cr}} = 2/d = 2$. Dashed line $P = P_{\mathrm{cr}}$ is the ($\mu-$ independent) value of $\mathcal{N}[\psi_\mu]$ for the case $V \equiv 0$. Solid curve is the curve $\mu \mapsto \mathcal{N}[\psi_\mu]$, where $\mu < \mu_\star = \inf(\sigma(-\Delta + V))$ for the case where V is a nontrivial periodic potential. Nontrivial effective mass matrix implies $P_\star = \lim_{\mu \uparrow \mu_\star} \mathcal{N}[\psi_\mu] < P_{\mathrm{cr}}$. See [59].

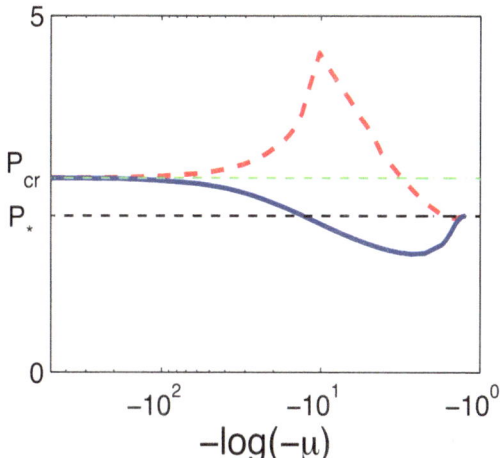

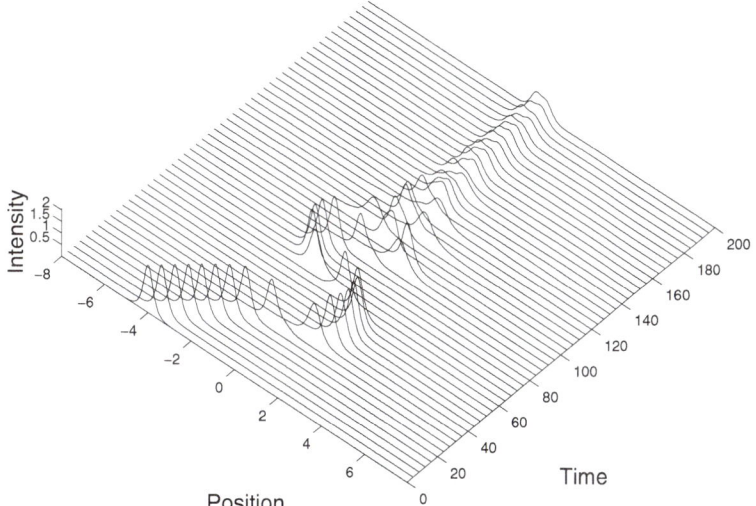

Figure 2.6 A soliton interacts with a defect, support around $x = 0$. As time advances, the coherent structure interacts with the defect. Part of the incident energy is scattered to infinity and part of it is trapped in the defect. The trapped energy settles down to a stable nonlinear defect mode.

with impurities, *e.g.* random defects in fabrication or those deliberately inserted into the medium to influence the propagation; see, for example, [48].

Referring to Figure 2.6 we sketch the different stages present in the dynamics of a soliton which is initially incident upon a potential well:

The results of formal asymptotic and numerical studies of soliton/defect interactions (see, for example, [11, 46, 48]) suggest a description of the time-dynamics in terms of overlapping time epochs. Very roughly, these are:

(i) short and long/intermediate time-scale *classical particle-like dynamics*, which govern motion and deformations of the soliton as it interacts with the defect and exchanges energy with the defect's internal modes (pinned nonlinear defect modes) and

(ii) the very long time scale, during which the system's energy resolves into outgoing solitons moving away from the defect, small amplitude waves which disperse to infinity and an asymptotically stable nonlinear ground state of supported in the defect. A key process in this *asymptotic resolution* is radiation damping due to coupling of discrete and continuum degrees of freedom. See sections 5 and 6.

Examples of rigorous analyses for *transient/intermediate time-scale* regimes:

(i) *Soliton scattering from a potential well*: Detailed reflection, transmission, and trapping for solitons incident on a potential barrier or potential well [21, 54–56].

(ii) *Soliton evolving in a potential well*: In [57] the evolution of an order one soliton in a potential well is considered. The detailed *nonlinear breathing dynamics*, as the soliton relaxes toward its asymptotic state are considered in [58]; see also section 6 and the related work [38, 40, 41].

In [82] the evolution of a weakly nonlinear (small amplitude) NLS/GP solutions in a double-well, for which there is symmetry breaking (see section 3.6) is studied. Figure 2.7 displays phase portrait of the reduced Hamiltonian dynamical system in [82]. Orbits around the left (respectively, right) equilibrium map to a soliton executing a long-time nearly periodic (back and forth) motion within the left (respectively, right) well of the double well. This analysis has been recently extended to certain orbits outside the separatrix [47].

Figure 2.7 Periodic dynamics of center of mass dynamics in the reduced (approximate) finite-dimensional Hamiltonian system [82]. Equilibria corresponding to stable asymmetric states centered on left and right local minima of double well. Oscillatory orbits decay to stable equilibria for the full, infinite-dimensional NLS/GP.

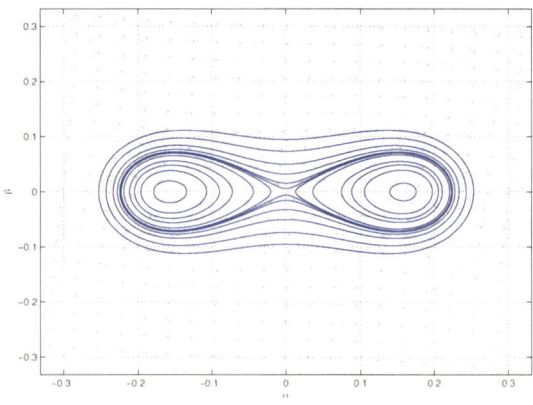

Figure 2.8 Numerically computed dynamics of center of mass in the NLS evolution [82]. Corresponding trajectory in the reduced phase portrait of Figure 2.7 cycles around several times outside the separatrix, crosses the separatrix, and then slowly spirals in toward the right stable equilibrium.

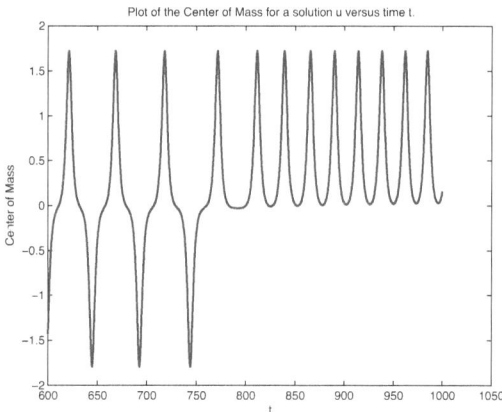

For $t \gg 1$ the center of mass eventually crystallizes on a stable nonlinear ground state. For $\mathcal{N} > \mathcal{N}_{cr}$ this asymptotic state will be an asymmetric state (section 3.6) centered in the left or in the right well. See Figure 2.8 which shows the computed center of mass motion for a solution of NLS/GP; initially, there is oscillatory motion among the left and right wells. However, during each cycle some of the soliton's energy radiated away to infinity. The corresponding motion in the reduced phase portrait (Figure 2.7) is damped and the actual center of mass trajectory is transverse to the level energy curves. Eventually the separatrix is crossed and the solution settles down to a stable asymmetric state [19, 98, 99]. We emphasize that this picture is only heuristic although there has been considerable progress toward understanding the radiation damping and ground state selection in such systems. We turn to these phenomena in the next section. We conclude this section noting work on the dynamics of forced/damped NLS and its reduced finite dimensional dynamics, capturing regular and chaotic behavior; see, for example, [10, 92].

5   Resonance, radiation damping and infinite time dynamics

Our next goal is to discuss results concerning the infinite time dynamics of NLS/GP. We begin with an informal discussion of the linear Schrödinger equation (NLS/GP with $g = 0$):

$$i\partial_t \psi \;=\; (-\Delta + V(x))\psi \tag{5.1}$$

Assume $V(x)$ is a smooth and sufficiently decaying real-valued potential for which $-\Delta + V$ has bound states :

$$(-\Delta + V)\psi_j = \Omega_j\psi_j, \psi_j \in L^2 j = 0, 1, \ldots, \psi_j \in L^2$$

with corresponding time harmonic solutions $\psi(x, t) = e^{-i\Omega_j t}\psi_j(x)$

Then, for sufficiently localized initial conditions, $\psi(0)$, the solution of the initial value problem for (5.1) may be written:

$$\psi(x, t) = e^{-i(-\Delta + V)t}\psi_0 = \sum_j \langle \psi_j, \psi(0) \rangle e^{-i\Omega_j t}\psi_j(x) + R(x, t) \qquad (5.2)$$

The sum in (5.2) is quasi-periodic while the second term, $R(x, t)$, radiates to zero in the sense that for appropriate $p > 2$ and $\alpha(p) > 0$

$$\|R(\cdot, t)\|_{L^p(\mathbb{R}^d)} \lesssim t^{-\alpha}, \text{ as } t \to \pm\infty \qquad (5.3)$$

Thus, as $t \to \infty$, the solution $\psi(x, t)$ tends to an asymptotic state which is quasi-periodic in time and localized in space.

Question: What is the large time behavior of solutions of the corresponding initial value problem for the nonlinear Schrödinger/Gross-Pitaevskii, $g \neq 0$?

For small amplitude it is natural to attempt expansion of solutions in the basis of localized eigenstates and continuum modes, associated with the unperturbed (solvable) linear Schrödinger equation. In terms of these coordinates, the original Hamiltonian PDE may be written as an equivalent dynamical system comprised of two weakly coupled subsystems:

- a finite or infinite dimensional subsystem with **discrete** degrees of freedom ("oscillators")
- an infinite dimensional system (wave equation) governing a **continuum** "field."

These two systems are coupled due to weak nonlinearity. If one "turns off" the nonlinear coupling there are localized in space and time-periodic solutions, corresponding the eigenstates of the linear Schrödinger equation. If nonlinearity is present, new frequencies are generated, and these may lead to resonances among discrete modes or between discrete and continuum radiation modes. The latter type of resonance plays a key role in understanding the radiation damping mechanism central to asymptotic relaxation of solutions to NLS/GP as $t \to \pm\infty$ and, in particular the phenomena of *ground state selection* and *energy equipartition*.

These results will be discussed in section 6. A variant was studied in the context of the nonlinear Klein Gordon equation in [4, 97]. In the following two subsections we present a very simple example of emergent effective damping in an infinite dimensional Hamiltonian problem.

5.1 Simple model - part 1: Resonant energy exchange between an oscillator and a wave-field

We consider a solvable toy model of an infinite dimensional Hamiltonian system comprised of two subsystems, one governing discrete and the other governing continuum degrees of freedom. We shall see that coupling of these subsystems leads to energy transfer from the discrete to the continuum modes and the emergence of effective damping. Our model is a variation on the work of Lamb (1900) [77] and Weisskopf-Wigner (1930) [113]; see also [14, 18, 68–70, 94–97]

Consider a system which couples an "oscillator" with amplitude $a(t)$ to a "field" with amplitude $u(x, t)$, $t \in \mathbb{R}$, $x \in \mathbb{R}$:

$$\frac{da^\varepsilon(t)}{dt} + i\omega a^\varepsilon(t) = -\varepsilon u^\varepsilon(0, t), \tag{5.4}$$

$$\partial_t u^\varepsilon(x, t) + c \, \partial_x u^\varepsilon(x, t) = \varepsilon \delta(x) a^\varepsilon(t). \tag{5.5}$$

Here, ε and c are taken to be a real parameters, say with $c > 0$, and $\delta(x)$ denotes the Dirac delta function. The amplitudes $a(t)$ and $u(x, t)$ are complex-valued.

Note that the dynamical system (5.4)–(5.5) conserves an energy, $\partial_t \mathcal{E}[a(t), u(\cdot, t)] = 0$.

$$\mathcal{E}[a(t), u(\cdot, t)] \equiv |a(t)|^2 + \int_{\mathbb{R}} |u(x, t)|^2 \, dx \tag{5.6}$$

We consider the initial value problem for (5.4)–(5.5) with initial data

$$a(0) \text{ arbitrary and } u(x, 0) = 0 \; ; \tag{5.7}$$

we perturb the oscillator initially, but not the field.

For $\varepsilon = 0$ the oscillator and wave-field are decoupled and there is an exact time-periodic and finite energy global solution, a *bound state*:

$$\begin{pmatrix} u^{\varepsilon=0}(x, t) \\ a^{\varepsilon=0}(t) \end{pmatrix} = e^{-i\omega t} \begin{pmatrix} 0 \\ a(0) \end{pmatrix}.$$

For $\varepsilon \neq 0$, the oscillator and field are coupled; as time evolves, energy can be transferred among the discrete and continuum degrees of freedom. It is simple to solve the initial value problem for any choice of ε and arbitrary initial data, e.g. by Laplace transform. In particular, one can proceed by solving the wave equation, (5.5), for $u^\varepsilon = u(x, t; a^\varepsilon)$ as a functional of discrete degree of freedom

$a^{\varepsilon}(t)$ and then substitute the resulting expression for $u^{\varepsilon}\,(0, t; a^{\varepsilon}(t))$ into the oscillator equation (5.4). The result is the closed oscillator

$$\frac{da^{\varepsilon}(t)}{dt} + i\omega a^{\varepsilon}(t) = -\frac{\varepsilon^2}{2c} a^{\varepsilon}(t), \quad t \geq 0. \tag{5.8}$$

with an effective damping term, whose solution is:

$$a^{\varepsilon}(t) = e^{-i\omega t}\, e^{-\frac{\varepsilon^2}{2c} t}\, a_0 = e^{-i(\omega - i\frac{\varepsilon^2}{2c})\, t}\, a_0, \quad t \geq 0. \tag{5.9}$$

That is, in terms of the oscillator, the *closed* infinite dimensional conserved system for $a^{\varepsilon}(t)$ and $u^{\varepsilon}(x, t)$ may equivalently be viewed in terms of a reduced *open and damped* finite dimensional system for $a^{\varepsilon}(t)$.

Solving (5.8) for $a^{\varepsilon}(t)$ and substituting into the expression for $u^{\varepsilon} = u(x, t; a^{\varepsilon})$, one obtains:

$$u^{\varepsilon}(x, t) = \begin{cases} 0, & x > ct \geq 0 \\ \frac{\varepsilon}{c}\, a_0\, e^{-i\omega(t-\frac{x}{c})}\, e^{-\frac{\varepsilon^2}{2c}(t-\frac{x}{c})}, & x < ct \end{cases}$$

To summarize, for $\varepsilon = 0$, the decoupled system has a time-periodic finite energy bound state. For the coupled system, $\varepsilon \neq 0$, the bound state has a finite lifetime; it decays on a timescale of order ε^{-2}. The bound state loses its energy to the continuum degrees of freedom; the lost energy is propagated to spatial infinity.

5.2 Simple model - part 2: Resonance, Effective damping, and Perturbations of Eigenvalues in Continuous Spectra

It's physically intuitive that an oscillator coupled to wave propagation in an infinite medium will damp. We now connect this damping to the classical notion of resonance.

We write the oscillator/wave-field model in the form

$$i\partial_t \psi = A(\lambda)\, \psi, \tag{5.10}$$

where

$$\psi(t) = \begin{pmatrix} u(x, t) \\ a(t) \end{pmatrix}, \quad A(\varepsilon) = \begin{pmatrix} -ic\partial_x & +i\varepsilon\delta(x) \\ -i\varepsilon\, \langle\delta(\cdot), \cdot\rangle & \omega \end{pmatrix} \tag{5.11}$$

Let

$$\psi(t) = e^{-iEt}\psi_0 = \begin{pmatrix} u(x,t) \\ a(t) \end{pmatrix} = e^{-iEt}\begin{pmatrix} u_0 \\ a_0 \end{pmatrix} \tag{5.12}$$

This yields the spectral problem

$$A(\varepsilon)\psi_0 = E\psi_0, \tag{5.13}$$

or equivalently

$$E\begin{pmatrix} u_0 \\ a_0 \end{pmatrix} = \begin{pmatrix} -ic\partial_x & +i\varepsilon\delta(x) \\ -i\varepsilon\langle\delta(y),\cdot\rangle & \omega \end{pmatrix}\begin{pmatrix} u_0 \\ a_0 \end{pmatrix}. \tag{5.14}$$

This spectral problem may be considered on the Hilbert space:

$$\mathcal{H} = \left\{ \begin{pmatrix} u \\ a \end{pmatrix} \in L^2(\mathbb{R}) \times \mathbb{C} : \|(u,a)\|_{\mathcal{H}} < \infty \right\}, \tag{5.15}$$

where

$$\|(u,a)\|_{\mathcal{H}} = \int_{\mathbb{R}} |U(x)|^2 dx + |a|^2 < \infty \tag{5.16}$$

We note that for $\varepsilon = 0$ (decoupling of oscillator and wave-field), we have

$$E\begin{pmatrix} u_0 \\ a_0 \end{pmatrix} = \begin{pmatrix} -ic\partial_x & 0 \\ 0 & \omega \end{pmatrix}\begin{pmatrix} u_0 \\ a_0 \end{pmatrix} \tag{5.17}$$

$A(\varepsilon = 0)$ is diagonal and has spectrum given by:

- a point eigenvalue at $E = \omega$ with corresponding eigenstate $\psi_\omega = \begin{pmatrix} 0 \\ 1 \end{pmatrix}$ and

- continuous spectrum $\mathbb{R} = \{E = ck : k \in \mathbb{R}\}$ with eigenstates $\psi_k = \begin{pmatrix} e^{ikx} \\ 0 \end{pmatrix}$, $k \in \mathbb{R}$.

Thus, $A(\varepsilon = 0)$ has an embedded eigenvalue, ω, in the continuous spectrum, $\mathbb{R}$, and the source the damping is seen to be this resonance. Understanding the coupling of oscillator and wave-field is therefore related to the perturbation theory of an embedded (non-isolated) point in the spectrum; see, for example, [53, 89, 95] and the references therein.

In this present simple setting, this perturbation problem can be treated as follows. Note that for $\varepsilon \neq 0$, since $c > 0$ we expect that energy is emitted by the oscillator and it gets carried to positive infinity through $u(x,t)$. Thus we expect that on a fixed compact set, $u(x,t)$, will decay to zero. Alternatively, if we choose $\alpha > 0$ and study

$U(x, t) = e^{-\alpha x} u(x, t)$, $U(x, t)$ can be expected to decay to zero as t advances. Indeed, it can be checked that $U(x, t)$ satisfies a *dissipative* PDE, for which the $L^2(\mathbb{R})$ norm of $U(x, t)$ decays as t increases; see the discussion of section 5.1.

This behavior is reflected in the spectral problem. Consider the change of variables $(u, a) \mapsto (e^{\alpha x} U, a)$ in the spectral problem (5.14). This yields

$$E \begin{pmatrix} U_0 \\ a_0 \end{pmatrix} = \begin{pmatrix} -ic\partial_x - ic\alpha & +i\varepsilon e^{-\alpha x}\delta(x) \\ -i\varepsilon \langle \delta(\cdot)e^{\alpha \cdot}, \cdot \rangle & \omega \end{pmatrix} \begin{pmatrix} U_0 \\ a_0 \end{pmatrix} = A_\alpha(\varepsilon) \begin{pmatrix} U_0 \\ a_0 \end{pmatrix}, \quad (5.18)$$

which when considered on $L^2(\mathbb{R})$ is equivalent to the spectral problem (5.14) for $(u(x), a)$ in the weighted function space:

$$\mathcal{H}_\alpha = \{(u(x), a) : \|u\|_{\mathcal{H}_\alpha} < \infty \}, , \quad (5.19)$$

where

$$\|u\|_{\mathcal{H}_\alpha} = \int_{\mathbb{R}} |u(x)|^2 e^{-2\alpha x} dx + |a|^2 < \infty, \quad \alpha > 0. \quad (5.20)$$

For $\varepsilon = 0$,

$$E \begin{pmatrix} v_0 \\ a_0 \end{pmatrix} = \begin{pmatrix} -ic\partial_x - ic\alpha & 0 \\ 0 & \omega \end{pmatrix} \begin{pmatrix} v_0 \\ a_0 \end{pmatrix} \quad (5.21)$$

$A_\alpha(0)$ is diagonal and has spectrum:

- point eigenvalue at $E = \omega$ with corresponding eigenstate $\psi_\omega = \begin{pmatrix} 0 \\ 1 \end{pmatrix}$
- continuous spectrum in the lower half plane along the horizontal line $\{E = ck - ic\alpha : k \in \mathbb{R}\}$ with eigenstates $\psi_k = \begin{pmatrix} e^{ikx} \\ 0 \end{pmatrix}$, $k \in \mathbb{R}$

Note that for $\alpha > 0$, ω is an *isolated* eigenvalue of $A_\alpha(\varepsilon = 0)$ and we can therefore implement a standard perturbation theory to calculate the effect of $\varepsilon \neq 0$ on this eigenvalue.

The first equation of (5.18) is

$$(-ic\partial_x - ic\alpha) v_0 + i\varepsilon\delta(x)e^{-\alpha x} a_0 = E v_0 \quad (5.22)$$

This implies for $x \neq 0$

$$v_0(x) = \begin{cases} e^{i(\frac{E}{c}x - \alpha)x} v_{0+} & x > 0 \\ e^{i(\frac{E}{c}x - \alpha)x} v_{0-} & x < 0 \end{cases}$$

Integration of (5.18) over a small neighborhood of $x = 0$ yields

$$v_{0+} - v_{0-} = +\frac{\varepsilon}{c}a_0 \qquad (5.23)$$

The second equation of (5.18) implies

$$(E - \omega)a_0 = -\frac{i\varepsilon}{2}\left(v_{0+} + v_{0-}\right) \qquad (5.24)$$

Now choose $v_{0+} \neq 0$ and $v_{0-} = 0$. Then,

$$E^{\varepsilon} = \omega - i\Gamma^{\varepsilon}, \quad \Gamma^{\varepsilon} = \frac{\varepsilon^2}{2c} > 0. \qquad (5.25)$$

The corresponding eigenstate is

$$v_{0,\alpha}^{\varepsilon}(x) = \begin{cases} +\frac{\varepsilon}{c}a_0\, e^{i\omega\frac{x}{c}}e^{\frac{1}{c}(\frac{\varepsilon^2}{2c}-\alpha c)x}, & x > 0 \\ 0, & x < 0 \end{cases}, \qquad (5.26)$$

which is in $L^2(\mathbb{R})$ provided $\alpha > \frac{\varepsilon^2}{2c^2}$.
For $\alpha = 0$, we have

$$v_0^{\varepsilon}(x) = \begin{cases} +\frac{\varepsilon}{c}a_0\, e^{i\omega\frac{x}{c}}e^{\frac{1}{c}\frac{\varepsilon^2}{2c}x}, & x > 0 \\ 0, & x < 0 \end{cases}. \qquad (5.27)$$

The solution (5.27) satisfies an *outgoing radiation condition* at $x = +\infty$ and is in $L^2(\mathbb{R}; e^{-2\alpha x}dx)$ provided $\alpha > \frac{\varepsilon^2}{2c^2}$..

To summarize: the $\varepsilon = 0$ eigenvalue problem has discrete eigenvalue, $E^0 = \omega$, embedded in the continuous spectrum, $\mathbb{R}$ with corresponding eigenstate $\begin{pmatrix} 0 \\ 1 \end{pmatrix}$. This eigenvalue perturbs, for $\varepsilon \neq 0$, to a complex energy in the lower half plane, $E^{\varepsilon} = \omega - i\Gamma^{\varepsilon}$, $\Gamma^{\varepsilon} > 0$ with corresponding eigenstate $\begin{pmatrix} v_0^{\varepsilon}(x) \\ 1 \end{pmatrix}$ which solves the eigenvalue equation with outgoing radiation condition at $x = +\infty$. The complex energy, E^{ε}, may be viewed as an eigenvalue with normalizable eigenstate, $v_0(x)$ in the weighted function space, $\mathcal{H}_\alpha$. Since E^{ε} is in the lower half plane, the corresponding time-dependent state is exponentially decaying as t increases.

Remark 5.1. *Finally, we remark that the complex energy E^{ε} may be viewed as a pole of the Green's function, when analytically continued from the upper half $E-$ plane to the lower half $E-$ plane.*

6   Ground state selection and energy equipartition in NLS/GP

In this section we return to the question raised in section 5, here restated. For simplicity we consider NLS/GP for the case with cubic nonlinearity ($\sigma = 1$):

$$i\partial_t \Phi = (-\Delta + V(x))\Phi + g|\Phi|^2\Phi, \quad x \in \mathbb{R}^3 \tag{6.1}$$

We assume that $-\Delta + V$ has <u>multiple</u> independent localized eigenstates. For simplicity we assume two distinct eigenvalues.

As noted, the linear time-evolution ($g = 0$), for $t \gg 1$, settles down to a quasi-periodic state, a linear superposition of time-periodic and spatially localized solutions, corresponding to the independent eigenstates. Note also that for $g \neq 0$ there may be multiple co-existing branches of nonlinear defect states; see, for example, the discussion of bound states for the case where V is a double-well potential, discussed in section 3.

Question: What is the long term ($t \uparrow \infty$) behavior of the NLS/GP ($g \neq 0$), (6.1) for initial data of small norm?

Our results show that under reasonable conditions (which have been explored experimentally [80, 99]) we have:

1. *Ground state selection [42, 43, 98, 99]:* The generic large time behavior of the initial value problem is periodic and is, in particular, a nonlinear ground state of the system. See also [7–9, 106, 107].
2. *Energy equipartition [43]:* For initial data conditions whose nonlinear ground state and excited state components are equal in L^2, the solution approaches a new nonlinear ground state, whose L^2 norm has gained an amount equal one-half that of the initial excited state energy. The other one-half of the excited state energy is radiated away.

To state our results precisely requires some mathematical setup. For simplicity we assume that the linear operator $-\Delta + V$ has the following properties:

(V1) V is real-valued and decays sufficiently rapidly, *e.g.* exponentially, as $|x|$ tends to infinity.
(V2)   The linear operator $-\Delta + V$ has two eigenvalues $E_{0\star} < E_{1\star} < 0$. $E_{0\star}$ is the lowest eigenvalue with ground state $\psi_{0\star} > 0$, the eigenvalue $E_{1\star}$ is possibly degenerate with multiplicity $N \geq 1$ and corresponding eigenvectors $\xi_{1\star}, \xi_{2\star}, \cdots, \xi_{N\star}$.
(V3)   Resonant coupling assumption

$$\omega_\star \equiv 2E_{1\star} - E_{0\star} > 0, . \tag{6.2}$$

We remark on (V3). An important role is played by the mechanism resonant coupling of discrete and continuum modes and energy transfer from localized states to dispersive radiation. In sections 5.1 and 5.2 we introduced and analyzed a simple model of this phenomenon. This model was linear and the resonance was due to a discrete eigenvalue, ω, embedded in the continuous spectrum. For NLS/GP, (6.1) resonant coupling of discrete and continuum modes arises due to higher harmonic generation by nonlinearity. The assumption (V3) is made to ensure that this coupling occurs at second order in (small) energy of the solution. However coupling at arbitrary higher order can be studied using the normal form ideas developed in [4, 16, 17, 39].

6.1 Linearization of NLS/GP about the ground state

Recall Theorem 3.2 on the family of nonlinear bound states, $E \mapsto \psi_E$, bifurcating from the ground state. We now consider the linearized NLS/GP, (6.1), time-dynamics about this family of solutions. Let

$$\Phi(x,t) = e^{-iEt} \left(\psi_E + u + iv \right),$$

where u and v are real and imaginary parts of the perturbation. Then the linearized perturbation equation about ψ_E is

$$\frac{\partial}{\partial t} \begin{pmatrix} u \\ v \end{pmatrix} = L(E) \begin{pmatrix} u \\ v \end{pmatrix} = JH(E) \begin{pmatrix} u \\ v \end{pmatrix}, \tag{6.3}$$

where

$$L(E) = \begin{pmatrix} 0 & L_-(E) \\ -L_+(E) & 0 \end{pmatrix} = \begin{pmatrix} 0 & 1 \\ -1 & 0 \end{pmatrix} \begin{pmatrix} L_+(E) & 0 \\ 0 & L_-(E) \end{pmatrix} \equiv JH(E). \tag{6.4}$$

The operators L_+ and L_- are given by:

$$L_-(E) = -\Delta - E + V + g(\psi_E)^2$$
$$L_+(E) = -\Delta - E + V + 3g(\psi_E)^2 \tag{6.5}$$

The following result on the linearized matrix-operator, $L(E)$, proved by standard perturbation theory [89], is given in [42] (Propositions 4.1 and 5.1):

Lemma 6.1. *Assume (V1), (V2), and (V3). Let $L(E)$ denote the linearized operator about a state on the branch of bifurcating bound states, given by Theorem 3.2:*

$$\psi_E(x) = \rho(E) \left(\psi_{0\star}(x) + \mathcal{O}\left(\rho(E)^2 \right) \right), \quad \text{where } \rho(E) \equiv |E_{0\star} - E|^{\frac{1}{2}} \left(|g| \int \psi_{0\star}^4 \right)^{-\frac{1}{2}} \tag{6.6}$$

For $E = E_{0\star}$, the matrix operator $L(E_{0\star})$ has complex conjugate eigenvalues $\pm i\beta_\star = \pm i(E_{1\star} - E_{0\star})$, each of multiplicity N. For $|E_{0\star} - E|$ and small, these perturb to (possibly degenerate) eigenvalues $\pm i\beta_1(E), \ldots, \pm i\beta_N(E)$ with corresponding neutral modes (eigenstates)

$$\begin{pmatrix} \xi_1 \\ \pm i\eta_1 \end{pmatrix}, \begin{pmatrix} \xi_2 \\ \pm i\eta_2 \end{pmatrix}, \cdots, \begin{pmatrix} \xi_N \\ \pm i\eta_N \end{pmatrix}$$

satisfying $\langle \xi_m, \eta_n \rangle = \delta_{m,n}, \ \langle \xi_m, \psi_E \rangle = \langle \eta_m, \partial_E \psi_E \rangle = 0 .$ (6.7)

Moreover,

$$0 \neq \lim_{E \to E_{0\star}} \xi_n = \lim_{E \to E_{0\star}} \eta_n \in \text{span}\{\xi_{1\star}, \xi_{2\star}, \ldots, \xi_{N\star}\}$$ (6.8)

in Sobolev H^k spaces for any $k > 0$.
Furthermore, we note that for $|E_{0\star} - E|$ sufficiently small

$$2\beta_n(E) + E \approx 2(E_{1\star} - E_{0\star}) + E_{0\star} = 2E_{1\star} - E_{0\star} > 0, \ n = 1, 2, \cdots, N,$$ (6.9)

Remark 6.1. *Equation* (6.9) *(see also* (V3)*,* (6.2)*) ensures coupling of discrete to continuous spectrum at second order in* $|E - E_{0\star}|$.

6.2 Ground state selection and energy equipartition

In this section we give a detailed description of the long-term evolution.

Theorem 6.1 (Ground State Selection). *Consider NLS/GP,* (6.1)*, with linear potential satisfying* (V1)*,* (V2)*, and* (V3)*, and cubic nonlinearity* ($\sigma = 1$)*. Assume that the non-negative (Fermi golden rule) expression for* $\Gamma_0(z, z^*)$ *in* (6.15) *is strictly positive. (See* [42, 43, 98] *for detailed statements, of all technical assumptions. This expression is always non-negative and generically strictly positive due to the assumption* (6.2)*.).*
Take initial conditions of the form:

$$\psi_0(x) = e^{i\gamma_0} \left[\psi_{E_0} + \alpha_0 \cdot \xi + i\beta_0 \cdot \eta + R_0 \right] ,$$ (6.10)

where γ_0 *and* $E_0 \in \mathcal{I} = (E_{0\star} - \delta_0, E_{0\star})$ *are real constants,* α_0 *and* β_0 *are real* $1 \times N$ *vectors, and* $R_0 : \mathbb{R}^3 \to \mathbb{C}$*, are such that*

$$|E_0 - E_{0\star}| \ll 1, \quad \frac{|\alpha_0|^2 + |\beta_0|^2}{\|\psi_{E_0}\|_2^2} \ll 1, \quad \|\langle x \rangle^4 R_0\|_{H^2} \lesssim |\alpha_0|^2 + |\beta_0|^2$$ (6.11)

Then there exist smooth functions $E(t) : \mathbb{R}^+ \to \mathcal{I}$, $\gamma(t) : \mathbb{R}^+ \to \mathbb{R}$, $z(t) : \mathbb{R}^+ \to \mathbb{C}^N$ and $R(x,t) : \mathbb{R}^3 \times \mathbb{R}^+ \to \mathbb{C}$ such that the solution of NLS/GP evolves in the form:

$$\psi(x,t) = e^{-i\int_0^t E(s)ds} e^{i\gamma(t)}$$

$$\times \left[\psi_{E(t)} + a_1(z,\bar{z})\, \partial_E \psi_E + i a_2(z,\bar{z})\, \psi_E + (\mathrm{Re}\, \tilde{z}[z,\bar{z}]) \cdot \xi \right.$$

$$\left. + i\,(\mathrm{Im}\tilde{z}[z,\bar{z}]) \cdot \eta + R \right], \tag{6.12}$$

where $\lim_{t\to\infty} E(t) = E_\infty$, for some $E_\infty \in \mathcal{I}$.
Here, $a_1(z,\bar{z})$, $a_2(z,\bar{z}) : \mathbb{C}^N \times \mathbb{C}^N \to \mathbb{R}$ and $\tilde{z} - z : \mathbb{C}^N \times \mathbb{C}^N \to \mathbb{C}^N$ are polynomials of z and $\bar{z}$, beginning with terms of order $|z|^2$.

(A) The dynamics of mass/energy transfer is captured by the following reduced dynamical system for the key modulating parameters, $E(t)$ and $z(t)$:

$$\frac{d}{dt}\, \|\psi_{E(t)}\|_2^2 = z^* \Gamma_0(z,\bar{z})\, z + \mathcal{S}_E(t), \tag{6.13}$$

$$\frac{d}{dt}\, |z(t)|^2 = -2z^* \Gamma_0(z,\bar{z})\, z + \mathcal{S}_z(t) . \tag{6.14}$$

Here, Γ_0 is a Fermi golden rule (damping) matrix given by [1]

$$\Gamma_0(z,z^*) \equiv g^2\, \Im \left\langle [-\Delta + V - \omega_\star - i0]^{-1}\, \psi_E\, (z \cdot \xi)^2,\ \psi_E\, (z \cdot \xi)^2 \right\rangle \geq c^2\, |z|^4 \tag{6.15}$$

and $\mathcal{S}_E(t)$, $\mathcal{S}_z(t) \lesssim (1+t)^{-p}$, with $\int_0^\infty \mathcal{S}_E(\tau)d\tau$ and $\int_0^\infty \mathcal{S}_z(\tau)d\tau = o(|z_0|)^2$.
(B) **Estimates on $z(t)$ and the correction $\mathbf{R}(t)$:** For all $t \geq 0$, we have $\|\mathbf{R}(t)\|_{H^2} \leq \epsilon_\infty$. Moreover, the following decay estimates hold:

$$\left\| (1 + x^2)^{-\nu} \mathbf{R}(t) \right\|_2 \leq C\, (\|\langle x \rangle^4 \psi_0 \|_{H^2})\, (1+t)^{-1}, \tag{6.16}$$

$$|z(t)| \leq C\, (\|\langle x \rangle^4 \psi_0 \|_{H^2})\, (1+t)^{-\frac{1}{2}} \tag{6.17}$$

Theorem 6.2 (Mass/Energy equipartition). *Consider NLS/GP, (6.1), under the technical hypotheses of Theorem 6.1 and assumptions on the initial data (6.11), where $|\alpha_0|^2 + |\beta_0|^2$ is a measure of the neutral modes' perturbation of the ground*

---

[1] $\Gamma_0(z, z^*)$ assumed to be strictly positive, is non-negative by the following:

$$\Im\, [-\Delta + V - \omega_\star - i0]^{-1} = \frac{1}{2i}\, \lim_{\delta \downarrow 0}\, \left([-\Delta + V - \omega_\star - i\delta]^{-1} - [-\Delta + V - \omega_\star + i\delta]^{-1} \right),$$

$$= \pi\, \delta\,(-\Delta + V - \omega_\star), \quad \text{where } \omega_\star \in \sigma_{cont}(-\Delta + V).$$

Note: $\Im\, [-\Delta + V - \omega_\star - i0]^{-1}$ projects onto the generalized mode at energy $\omega_\star \in \sigma_{cont}(-\Delta + V)$.

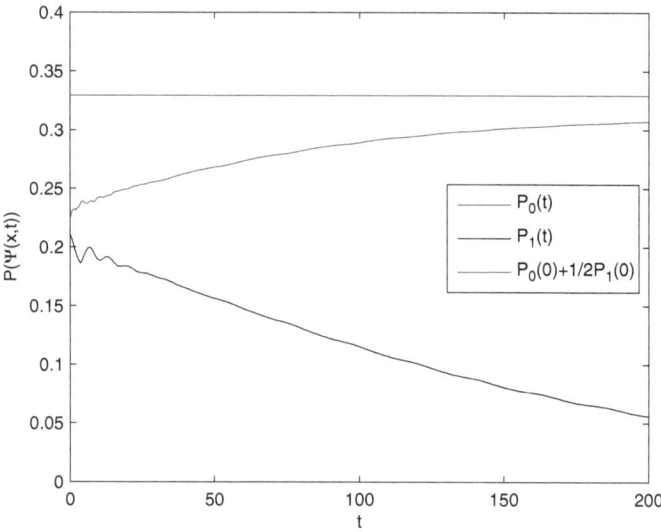

Figure 2.9 Numerical verification ground state selection and asymptotic energy equipartition. Computations performed by E. Shlizerman.

state. Recall that E_0 is the energy of the initial nonlinear ground state and E_∞ is the asymptotic energy, guaranteed by Theorem 6.1. Then, in the limit as $t \to \infty$, one half of the neutral modes' mass contributes to forming a more massive asymptotic ground state and one half is radiated away as dispersive waves:

$$\|\psi_{E_\infty}\|_2^2 = \|\psi_{E_0}\|_2^2 + \frac{1}{2} \left[|\alpha_0|^2 + |\beta_0|^2 \right] + o \left(|\alpha_0|^2 + |\beta_0|^2 \right). \tag{6.18}$$

Figure 2.9 illustrates the phenomena of ground state selection and mass/energy equipartition in NLS/GP. These simulations are for the case of double-well potentials (see section 3.6). Asymptotic energy equipartition has been verified for the case when the stable ground state is symmetric ($\mathcal{N} < \mathcal{N}_{cr}$) and the case when the stable ground states is asymmetric ($\mathcal{N} > \mathcal{N}_{cr}$).

A result on equipartition of energy holds as well, for NLS-GP with a general power nonlinearity, (2.1), under more restrictive hypotheses [43]. Finally, we remark that it would be of interest to establish detailed results on energy-transfer in systems with multiple bound states for subcritical nonlinearities. In this setting the current perturbative treatment of small amplitude dispersive waves does not apply. A step in this direction is in the work [67, 71, 72], where asymptotic stability for systems with a single family of nonlinear bound states is treated by a novel time-dependent linearization procedure.

7 A nonlinear toy model of nonlinearity-induced energy transfer

In this section we explain the idea behind the results on ground state selection and energy equipartition. In the style of sections 5.1 and 5.2 we introduce a toy minimal model, here nonlinear, which captures the essential mechanisms and is, by comparison, simpler to analyze.

Our model is for the interaction of three amplitudes:

1. a "ground state" complex amplitude, $\alpha_0(t)$ with associated frequency $E_{0\star}$
2. an "excited state" complex amplitude, $\alpha_1(t)$ with associated frequency $E_{1\star}$, and
3. a "continuum wave-field" complex amplitude, $R(x,t)$, with spectrum of frequencies given by the positive real-line $[0, \infty)$.

As with NLS/GP, our model has cubic nonlinearity and the assumption

$$\omega_\star \equiv 2E_{1\star} - E_{0\star} > 0 \tag{7.1}$$

is assumed to assure coupling of discrete and continuum modes at second order in the solution amplitude. We also introduce a function, $\chi(x)$, sufficiently rapidly decaying at spatial infinity. (Recall that in section 5.1 we took $\chi(x) = \delta(x)$.)

Our nonlinear model is the following:

$$i\partial_t \alpha_0(t) - \Omega_0 \alpha_0 = g \left\langle \chi, \overline{R(t)} \right\rangle \alpha_1^2(t) \tag{7.2}$$

$$i\partial_t \alpha_1(t) - \Omega_1 \alpha_1 = 2g \left\langle \chi, R(t) \right\rangle \alpha_0(t) \overline{\alpha_1(t)} \tag{7.3}$$

$$i\partial_t R(x,t) = -\Delta R(x,t) + g \chi(x) \alpha_1^2(t) \overline{\alpha_0(t)} \tag{7.4}$$

A first observation is that the system (7.2)–(7.4) is Hamiltonian and has the time-invariant quantity

$$\frac{d}{dt} \left(|\alpha_0(t)|^2 + |\alpha_1(t)|^2 + \int_{\mathbb{R}^d} |R(x,t)|^2 \, dx \right) = 0$$

To solve (7.2)–(7.4), we first separate fast and slow scales by introducing slowly varying amplitudes, A_0 and A_1:

$$\alpha_0(t) = e^{-i\Omega_0 t} A_0(t), \quad \alpha_1(t) = e^{-i\Omega_1 t} A_1(t)$$

Then, the system for A_0, A_1 and R becomes:

$$i\partial_t A_0(t) = g \left\langle \chi, \overline{R(t)} \right\rangle e^{-i\omega_\star t} A_1^2(t) \tag{7.5}$$

$$i\partial_t A_1(t) = 2g \left\langle \chi, R(t) \right\rangle e^{i\omega_\star t} A_0(t) \overline{A_1(t)} \tag{7.6}$$

$$i\partial_t R(x,t) = -\Delta R(x,t) + g \chi(x) e^{-i\omega_\star t} A_1^2(t) \overline{A_0(t)} \tag{7.7}$$

Note that since the spectrum of $-\Delta = [0, \infty)$ and since $\omega_\star > 0$ (by assumption) the Schrödinger wave equation (7.7) is resonantly forced. We proceed now to solve for $R(x, t)$, keeping its dominant contributions. Duhamel's principle and a regularization, motivated by the local decay estimate (7.9) below, gives:

$$R(t) = e^{i\Delta t} R(0) - ig\, e^{i\Delta t} \int_0^t e^{-i(-\Delta + \omega_\star)s}\, \chi \overline{A_0(s)}\, A_1^2(s)\, ds$$

$$\overset{\delta \downarrow 0}{=} e^{i\Delta t} R(0) - ig\, e^{i\Delta t} \int_0^t \frac{1}{i(-\Delta + \omega_\star - i\delta)} \frac{d}{ds} \left[e^{i(-\Delta - \omega_\star)s} \right] \chi \overline{A_0(s)} A_1^2(s)\, ds$$

Next we have, by integration by parts,

$$R(x, t) \sim e^{i\Delta t} R(0) - g e^{-i\omega_\star t} \frac{1}{-\Delta - \omega_\star - i0}\, \chi\, \overline{A_0(t)} A_1^2(t)$$

$$+ \overline{A_0(0)} A_1^2(0)\, \frac{e^{i\Delta t}}{-\Delta - \omega_\star - i0}\, \chi$$

$$+ \int_0^t \frac{e^{i\Delta(t-s)}}{-\Delta - \omega_\star - i0}\, \chi\, \frac{d}{ds} \left[\overline{A_0(s)} A_1^2(s) \right]\, ds\ ,$$

which may be written as

$$R(t) \sim -g e^{-i\omega_\star t}\, \frac{1}{-\Delta - \omega_\star - i0}\, \chi\, \overline{A_0(t)} A_1^2(t) + \mathcal{O}\left(t^{-\frac{3}{2}} \right) \tag{7.8}$$

where the error term is obtained using the *local energy decay estimate*:

$$\left\| \langle x \rangle^{-\nu}\, \frac{e^{i\Delta t}}{-\Delta - \omega_\star - i0}\, \langle x \rangle^{-\nu} \right\|_{L^2(\mathbb{R}^3) \to L^2(\mathbb{R}^3)} = \mathcal{O}(t^{-\frac{3}{2}}), \quad t \to +\infty, \tag{7.9}$$

and where ν is sufficiently large and positive. Substitution of the leading order terms of $R(t)$ in the expansion (7.8) into (7.5)–(7.6) and use of the distributional identity

$$(-\Delta - \omega_\star - i0)^{-1} = \text{P.V.}\, \frac{1}{-\Delta - \omega_\star} + i\pi\, \delta(-\Delta - \omega_\star)$$

yields the dissipative system for the amplitudes A_0 and A_1:

$$\partial_t A_0 \approx +\Gamma\, |A_1|^4 A_0\ , \quad \partial_t A_1 \approx -2\Gamma\, |A_0|^2 |A_1|^2 A_1 \tag{7.10}$$

where $\Gamma \sim g^2\, |\hat{\chi}(\omega_\star)|^2 \geq 0$ and is generically strictly positive. This is the analogue of the strict positivity discussed in the statement of Theorem 6.2. Here, $\hat{\chi}$ denotes the Fourier transform of χ.

The omitted correction terms in (7.10) can be controlled in terms of the quantities $\|R(t)\|_{L^p}$ ($p > 2$ and sufficiently large), $\|\langle x \rangle^{-\nu} R(t)\|_{L^2}$ and $\| \langle x \rangle^{-\nu} e^{i\Delta t} (-\Delta - \omega_* - i0)^{-1} \langle x \rangle^{-\nu} \|_{L^2 \to L^2}$.

Finally, ground state selection with energy equipartition can be seen through the reduction (7.10). Indeed, setting $P_0 = |\alpha_0|^2$, $P_1 = |\alpha_1|^2$ we obtain

$$\frac{dP_0}{dt} \sim \Gamma P_1^2 P_0 , \qquad \frac{dP_1}{dt} \sim 2\Gamma P_1^2 P_0$$

It follows that $2P_0(t) + P_1(t) \sim 2P_0(0) + P_1(0)$. Sending $t \to \infty$ and using that $P_1(t) \to 0$, (indeed, $\Gamma > 0$), we obtain $P_0(\infty) = P_0(0) + \frac{1}{2}P_1(0)$.

8 Concluding remarks

A fundamental question in the theory of nonlinear waves for non-integrable PDEs is whether and in what sense arbitrary finite-energy initial conditions evolve toward the family of available nonlinear bound states and radiation. This question is often referred to as the *Soliton Resolution Conjecture*; see, for example, [105]. The asymptotic stability/nonlinear scattering results of the type discussed in section 6, being based on normal forms ideas, spectral theory of linearized operators, dispersive estimates and low-energy (perturbative) scattering methods, give a local picture of the phase space near families of nonlinear bound states. In contrast, integrable nonlinear systems which can be mapped to exactly solvable linear motions allow for a global description of the dynamics [20, 22–24]. We also note the important line of research by Merle *et al.* which is based on the monotone evolution properties of appropriately designed local energies (see, for example, in [81, 83]) and does not rely on dispersive time-decay estimates of the linearized flow. It would be of great interest to adapt these methods for use in tandem with dynamical systems ideas to situations where the background linear medium, *i.e.* a potential $V(x)$, gives rise to structures (defect modes) which interact with "free solitons." A recent result, related to this point, is [75].

With a view toward understanding asymptotic resolution in non-integrable Hamiltonian PDEs, one of the simplest global questions to ask is the following. Let V denote a smooth, radially symmetric, and uni-modal potential well which decays very rapidly at spatial infinity. Assume further that $-\Delta + V$ has a finite number of bound states. Consider NLS/GP with a repulsive nonlinear potential ($g = +1$) with radially symmetric initial conditions:

$$i\partial_t \Psi = -\Delta \Psi + V(x)\Psi + |\Psi|^{p-1}\Psi, \quad \Psi(x,0) = \Psi_0(|x|) \in H^1_{radial}(\mathbb{R}^d)$$

It is natural to expect that energy which remains spatially localized must be captured by the family of nonlinear defect modes, *i.e.* bound states, localized within the potential well, which lie on solution branches bifurcating from the linear bound states of $-\Delta + V$ [90]; see section 3. Indeed, any concentration of energy outside the well should disperse to zero, since for $|x|$ large the equation is a translation invariant NLS equation with repulsive (defocusing) nonlinearity having no bound states. We conjecture that for arbitrary initial data solutions either disperse to zero or approach a nonlinear defect state, and furthermore that generic initial conditions approach a stable nonlinear defect state.

A step in this direction is the result of Tao [104] for $1 + \frac{4}{d} < p < 2^\star$, in very high spatial dimensions. *For spatial dimensions $d \geq 11$ there is a compact attractor. That is, there exists K, a compact subset of $H^1_{radial}(\mathbb{R}^d)$, such that K is invariant under the NLS/GP flow. Moreover, if $u \in C^0_t H^1_x(\mathbb{R} \times \mathbb{R}^d)$ is a global-in-time solution of NLS/GP, then there exists $u_+ \in H^1(\mathbb{R}^d_{radial})$ such that $dist_{H^1}\left(\Psi(t) - e^{i\Delta t}u_+, K\right) \to 0$, as $t \to +\infty$.*
It is natural to conjecture that K is the set of nonlinear bound (defect) states. See also [65]

Finally we mention that the detailed dynamical picture of energy transfer from discrete to radiation modes has connections with important questions in applied physics. Note that the Fermi golden rule damping matrix, $\Gamma_0(z, z)$, appearing in Theorems 6.1 and 6.2, which controls the rate of decay of excited states (neutral modes) is controlled by the density of states of the linearized operator near the resonant frequency $\omega_\star$; see (6.2).
Control problem: Can one design the potential, $V(x)$, in NLS/GP in order to inhibit/enhance energy transfer and relaxation (decay) to the system's asymptotic state?

This relates to the problem of controlling the spontaneous emission rate of atoms by modifying the background environment of the atom (for us, the background linear potential), thereby controlling the density of states [13]. A problem of this type was investigated analytically and computationally for a closely related parametrically forced linear Schrödinger equation [85]. Such a study for NLS/GP could inform the design of experiments, such as those reported on in [80].

Acknowledgements The author would like to thank J. Marzuola and E. Shlizerman for stimulating discussions and helpful comments on this chapter. He also wishes to thank the referees for their careful reading and suggested improvements. This work was supported, in part, by NSF Grants DMS-1008855 and DMS-1412560, and a grant from the Simons Foundation (#376319).

References

1. G. P. Agrawal. *Nonlinear Fiber Optics*. Optics and Photonics. Academic Press, 2nd edition, 1995.
2. W.H. Aschbacher, J. Fröhlich, G.M. Graf, K. Schnee, and M. Troyer. Symmetry breaking regime in the nonlinear Hartree equation. *J. Math. Phys.*, 43:3879–3891, 2002.
3. N. W. Ashcroft and N. D. Mermin. *Solid State Physics*. Saunders College Publishing, Harcourt Brace Publishers, 1976.

4. D. Bambusi and C. Cuccagna. On dispersion of small energy solutions of the nonlinear Klein Gordon equation with a potential. *Am. J. Math.*, 133(5):1421–1468, 2011.
5. H. Berestycki and P.-L. Lions. Nonlinear scalar field equations, . Existence of a ground state. *Arch. Rat. Mech. Anal.*, 1983.
6. J. Bourgain. *Global Solutions of Nonlinear Schrödinger Equations*, volume 46 of *Colloquium Publications*. AMS, 1999.
7. V. S. Buslaev and G. Perel'man. *On the stability of solitary waves for nonlinear Schrödinger equations*, volume 164 of *Amer. Math. Soc. Transl. Ser. 2*. Amer. Math. Soc., Providence, RI, 1995.
8. V. S. Buslaev and G. S. Perelman. Scattering for the nonlinear Schrödinger equation: states that are close to a soliton. *St. Petersburg Math. J.*, 4(6):1111–1142, 1993.
9. V. S. Buslaev and C. Sulem. On asymptotic stability of solitary waves for nonlinear Schrödinger equations. *Ann. Inst. H. Poincaré Anal. Non Linéaire*, 20(3):419–475, 2003.
10. D. Cai, D.W. McLaughlin, and K.T.R. McLaughlin. The nonlinear Schrödinger equation as both a PDE and a dynamical system. In *Handbook of Dynamical Systems*, volume 2, pages 599–675. North-Holland, 2002.
11. X. D. Cao and B. A. Malomed. Soliton-defect collisions in the nonlinear schrodinger equation. *Physics Letters A*, 206:177–182, 1995.
12. T. Cazenave and P.L. Lions. Orbital stability of standing waves for some nonlinear Schrödinger equations. *Comm. Math. Phys.*, 85:549–561, 1982.
13. C. Cohen-Tannoudji, J. Dupont-Roc, and G. Grynberg. *Atom-Photon Interactions - Basic Processes and Applications*. Wiley-Interscience, 1992.
14. O. Costin and A. Soffer. Resonance theory for Schrödinger operators. *Comm. Math. Phys.*, 224:133–152, 2001.
15. R. Courant and D. Hilbert. *Methods of Mathematical Physics*. Interscience Publishers, Inc., New York, N.Y., 1953.
16. S. Cuccagna. The Hamiltonian structure of the nonlinear Schrödinger equation and the asymptotic stability of its ground states. *Comm. Math. Phys.*, 305(2):279–331, 2011.
17. S. Cuccagna. On the Darboux and Birkhoff steps in the asymptotic stability of solitons. *Rend. Istit. Mat. Univ. Trieste*, 44:197–257, 2012.
18. S. Cuccagna, E. Kirr, and D. Pelinovsky. Parametric resonance of ground states in the nonlinear Schrödinger equation. *J. Differential Equations*, 220(1):85–120, 2006.
19. S. Cuccagna and J. Marzuola. On instability for the quintic nonlinear schrödinger equation of some approximate periodic solutions. *Indiana J. Math.*, 61(6):2053–2083, 2013.
20. S. Cuccagna and D. Pelinovsky. The asymptotic stability of solitons in the cubic NLS equation on the line. *Applicable Analysis*, 93(791–822), 2014.
21. K. Datchen and J. Holmer. Fast soliton scattering by attractive delta impurities. *Commun. Partial Differential Equations*, 34:1074–1113, 2009.
22. P. Deift and J. Park. Long-time asymptotics for solutions of the NLS equation with a delta potential and even initial data. *Int. Math. Res. Not. IMRN*, 24:5505–5624, 2011.
23. P.A. Deift, A.R. Its, and X. Zhou. Long-time asymptotics for integrable nonlinear wave equations,. In *Important developments in soliton theory*, Springer Ser. Nonlinear Dynam., pages 181–204. Springer, Berlin, 1993.
24. P.A. Deift and X. Zhou. Long-time asymptotics for solutions of the nls equation with initialdata in weighted sobolev spaces. *Comm. Pure Appl. Math.*, 56:1029–1077, 2003.
25. T. Dohnal, D. Pelinovsky, and G. Schneider. Coupled-mode equations and gap solitons in a two-dimensional nonlinear elliptic problem with a separable potential. *J. Nonlin. Sci.*, 19:95–131, 2009.
26. T. Dohnal and H. Uecker. Coupled-mode equations and gap solitons for he 2d Gross-Pitaevskii equation with a non-separable periodic potential. *Physica D*, 238:860–879, 2009.
27. V. Duchêne, J.L. Marzuola, and M.I. Weinstein. Wave operator bounds for one-dimensional schrödinger operators with singular potentials and applications. *J. Math. Phys.*, 52:013505, 2011.

28. V. Duchene, I. Vukicevic, and M.I. Weinstein. Oscillatory and localized perturbations of periodic structures and the bifurcation of defect modes. http://arxiv.org/abs/1407.8403, 2014.

29. V. Duchene, I. Vukicevic, and M.I. Weinstein. Homogenized description of defect modes in periodic structures with localized defects. *Communications in Mathematical Sciences*, 2015.

30. L. Erdös, B. Schlein, and H.T. Yau. Derivation of the cubic non-linear Schrödinger equation from quantum dynamics of many-body systems. *Invent. Math.*, 59(12):1659–1741, 2007.

31. C. L. Fefferman and M. I. Weinstein. Wave packets in honeycomb lattice structures and two-dimensional Dirac equations. *Commun. Math. Phys.*, 326:251–286, 2014.

32. C.L. Fefferman, J.P. Lee-Thorp, and M.I. Weinstein. Topologically protected states in one-dimensional continuous systems. *Proc. Nat. Acad. Sci.*, 111(24):8759–8763, 2014.

33. C.L. Fefferman, J.P. Lee-Thorp, and M.I. Weinstein. Topologically protected states in one-dimensional systems. *J. American Math. Soc.*, http://arxiv-web.arxiv.org/abs/1405.4569, 2014.

34. C.L. Fefferman and M.I. Weinstein. Waves in honeycomb structures. In D. Lannes, editor, *Proceedings of Journees Equations aux derivees partialles, Biarretz, 3-7 juin 2012*, volume GDR 243 4 (CNRS), – http://arxiv.org/abs/1212.6684, 2012.

35. C.L. Fefferman, J.P. Lee-Thorp, and M.I. Weinstein. Edge states in honeycomb structures, http://arxiv.org/abs/1506.06111, submitted.

36. G. Fibich. Some modern aspects of self-focusing theory. In Y.R. Shen R.W. Boyd, S.G. Lukishova, editor, *Self-Focusing: Past and Present*. Springer, 2009.

37. G. Fibich. *The Nonlinear Schrödinger Equation: Singularity Solutions and Optical Collapse*, volume 192 of *Applied Mathematical Sciences*. Springer-Verlag, 2014.

38. J. Fröhlich, S. Gustafson, B. L. G. Jonsson, and I. M. Sigal. Solitary wave dynamics in an external potential. *Comm. Math. Phys.*, 250(3):613–642, 2004.

39. Z. Gang. Perturbation expansion and Nth order Fermi golden rule of the nonlinear Schrödinger equations. *J. Math. Phys.*, 48(5):053509, 23, 2007.

40. Z. Gang and I. M. Sigal. On soliton dynamics in nonlinear Schrödinger equations. *Geom. Funct. Anal.*, 16(6):1377–1390, 2006.

41. Z. Gang and I.M. Sigal. Relaxation of solitons in nonlinear Schrödinger equations with potential. *Adv. Math.*, 216(2):443–490, 2007.

42. Z. Gang and M.I. Weinstein. Dynamics of nonlinear Schrödinger – Gross-Pitaevskii equations; mass transfer in systems with solitons and degenerate neutral modes. *Analysis and PDE*, 1(3), 2008.

43. Z. Gang and M.I. Weinstein. Equipartition of energy in Nonlinear Schrödinger / Gross-Pitaevskii Equations. *Applied Math. Research Express (AMRX)*, 2011.

44. B. Gidas, W.-M. Ni, and L. Nirenberg. Symmetry and related properties via the maximum principle. *Comm. Math. Phys.*, 68(3):209–243, 1979.

45. M. Golubitsky, I. Stewart, and D. Schaefer. *Singularities and Groups in Bifurcation Theory - Volume 2*. Springer-Verlag, 1988.

46. R.H. Goodman, P.J. Holmes, and M.I. Weinstein. Strong NLS soliton-defect interactions. *Physica D*, 161:21–44, 2004.

47. R.H. Goodman, J. Marzuola, and M.I. Weinstein. Self-trapping and josephson tunneling solutions to the nonlinear Schrödinger – Gross-Pitaevskii equation. *Discrete and Continuous Dynamical Systems - A*, 35(1):225–246, 2015.

48. R.H. Goodman, R.E. Slusher, and M.I. Weinstein. Stopping light on a defect. *J. Opt. Soc. B*, 19(7):1635–1652, 2002.

49. M. Grillakis, M. Machedon, and D. Margetis. Second order corrections to weakly interacting bosons, i. *Communications in Mathematical Physics*, 294(1):273–301, 2010.

50. M.G. Grillakis. Analysis of the linearization around a critical point of an infinite-dimensional hamiltonian system. *Comm. Pure Appl. Math.*, 43(3):299–333, 1990.

51. M.G. Grillakis, J. Shatah, and W.A. Strauss. Stability theory of solitary waves in the presence of symmetry. I. *J. Func. Anal.*, 74(1):160–197, 1987.

52. E. M. Harrell. Double wells. *Comm. Math. Phys.*, 75:239–261, 1980.

53. P. D. Hislop and I. M. Sigal. *Introduction to Spectral Theory: With applications to Schrödinger Operators*, volume 113 of *Applied Mathematical Sciences*. Springer, 1996.
54. J. Holmer, J. Marzuola, and M. Zworski. Fast soliton scattering by delta impurities. *Comm. Math. Phys.*, 274:349–367, 2007.
55. J. Holmer, J. Marzuola, and M. Zworski. Soliton splitting by delta impurities. *J. Nonlinear Sci.*, 7:349–367, 2007.
56. J. Holmer and M. Zworski. Slow soliton interaction with delta impurities. *J. Mod. Dyn.*, 1(4):689–718, 2007.
57. J. Holmer and M. Zworski. Soliton interaction with slowly varying potentials. *IMRN Internat. Math. Res. Notices*, 2008.
58. J. Holmer and M. Zworski. Breathing patterns in nonlinear relaxation. *Nonlinearity*, 22:1259–1301, 2009.
59. B. Ilan and M. I. Weinstein. Band edge solitons, nonlinear Schroedinger / Gross-Pitaevskii equations and effective media. *Multiscale Model. and Simul.*, 8(4):1055–1101, 2010.
60. R.K. Jackson and M.I. Weinstein. Geometric analysis of bifurcation and symmetry breaking in a Gross-Pitaevskii equation. *J. Stat. Phys.*, 116:881–905, 2004.
61. M. Jenkinson and M.I. Weinstein. On-site and off-site solitary waves of the discrete nonlinear Schroedinger equation in multiple dimensions. http://arxiv.org/abs/1405.3892, 2014.
62. C.K.R.T. Jones. An instability mechanism for radially symmetric standing waves of a nonlinear Schrödinger equation. *J. Diff. Eqns*, 71(1):34–62, 1988.
63. T. Kapitula, P. G. Kevrekidis, and Z. Chen. Three is a crowd: Solitary waves in photorefractive media with three potential wells. *SIAM Journal of Applied Dynamical Systems*, 5:598–633, 2006.
64. T. Kato. On nonlinear Schrödinger equations. *Ann. Inst. H. Poincaré Phys. Theor.*, 46:113–129, 1987.
65. C. Kenig, The concentration-compactness rigidity method for critical dispersive and wave equations. In Nonlinear Partial Differential Equations. Springer Basel pp. 117–149, 2012.
66. E. Kirr, P. G. Kevrekidis, and D. E. Pelinovsky. Symmetry breaking in the nonlinear Schrödinger equation with a symmetric potential. *Comm. Math. Phys.*, 308(3):795–844, 2011.
67. E. Kirr and O. Mizrak. Asymptotic stability of ground states in 3D nonlinear Schrödinger equation including subcritical cases. *J. Func. Anal.*, 257(12):3691–3747, 2009.
68. E. Kirr and M.I. Weinstein. Parametrically excited Hamiltonian partial differential equations. *SIAM. J. Math. Anal.*, 33(1):16–52, 2001.
69. E. Kirr and M.I. Weinstein. Metastable states in parametrically excited multimode Hamiltonian systems. *Commun. Math. Phys.*, 236(2):335–372, 2003.
70. E. Kirr and M.I. Weinstein. Diffusion of power in randomly perturbed hamiltonian partial differential equations. *Comm. Math. Phys.*, 255(2):293–328, 2005.
71. E. Kirr and A. Zarnescu. On the asymptotic stability of bound states in 2D cubic Schrödinger equation. *Comm. Math. Phys.*, 272(2):443–468, 2007.
72. E. Kirr and A. Zarnescu. Asymptotic stability of ground states in 2D nonlinear Schrödinger equation including subcritical cases. *J. Differential Equations*, 247(3):710–735, 2009.
73. E.W. Kirr, P. G. Kevrekidis, E. Shlizerman, and M.I. Weinstein. Symmetry breaking in the nonlinear Schrödinger / Gross-Pitaevskii equation. *SIAM J. Math. Anal.*, 40(2):566–604, 2008.
74. W. Kirsch and B. Simon. Comparison theorems for the gap of Schrödinger operators. *J. Func. Anal.*, 75:396–410, 1987.
75. M. Kowalczyk, Y. Martel and C. Munoz, Kink dynamics in the ϕ^4 model: asymptotic stability for odd perturbations in the energy space, arxiv.org/pdf/1506.07420.pdf
76. M. K. Kwong. Uniqueness of positive solutions of $\Delta u - u + u^p = 0$ in R^n. *Arch. Rat. Mech. Anal.*, 105:243–266, 1989.
77. H. Lamb. On a peculiarity of the wave-system due to the free vibrations of a nucleus in an extended medium. *Proc. London Math. Soc.*, 32:208–211, 1900.
78. E.H. Lieb and M. Loss. *Analysis*, volume 14 of *Graduate Studies in Mathematics*. AMS, Providence, Rhode Island, 2nd edition, 1997.

79. P.-L. Lions. The concentration compactness principle in the calculus of variations i,ii. *Anal. I.H.P. Anal. Nonlin.*, 1:109–145, 223–283, 1984.

80. D. Mandelik, Y. Lahini, and Y. Silberberg. Nonlinearly induced relaxation to the ground state in a two-level system. *Phys. Rev. Lett.*, 95:073902, 2005.

81. Y. Martel and F. Merle. Asymptotic stability of solitons for subcritical generalized KdV equations. *Arch. Rat. Mech. Anal.*, 157:219–254, 2001.

82. J. Marzuola and M.I. Weinstein. Long time dynamics near the symmetry breaking bifurcation for nonlinear Schrödinger / Gross-Pitaevskii equation. *Discrete Contin. Dyn. Syst. A*, 28(4):1505–1554, 2010.

83. F. Merle and P. Raphael. On the universality of blow-up profile for the l^2 critical nonlinear Schrödinger equation. *Invent. Math.*, 156(565–672), 2004.

84. L. Nirenberg. *Topics in Nonlinear Functional Analysis*, volume 6 of *Courant Institute Lecture Notes*. American Mathematical Society, Providence, Rhode Island, 1974.

85. B. Osting and M.I. Weinstein. Emergence of periodic structure from maximizing the lifetime of a bound state coupled to radiation. *SIAM J. Multiscale Model. and Simul.*, 9(2):654–685, 2011.

86. D. E. Pelinovsky, A. A. Sukhorukov, and Y. S. Kivshar. Bifurcations and stability of gap solitons in periodic potentials. *Phys. Rev. E*, 70:036618, 2004.

87. C.A. Pillet and C.E. Wayne. Invariant manifolds for a class of dispersive Hamiltonian partial differential equations. *J. Diff. Eqns*, 1997.

88. L.P. Pitaevskii and S. Stringari. *Bose-Einstein Condensation*, volume 116 of *International Series of Monographs on Physics*. Clarendon Press, 2003.

89. M. Reed and B. Simon. *Modern Methods of Mathematical Physics, IV*. Academic Press, 1978.

90. H. A. Rose and M. I. Weinstein. On the bound states of the nonlinear Schrödinger equation with a linear potential. *Physica D*, 30:207–218, 1988.

91. Z. Shi and J. Yang. Solitary waves bifurcated from Bloch-band edges in two-dimensional periodic media. *Phys. Rev. E*, 75:056602, 2007.

92. E. Shlizerman and V. Rom-Kedar. Classification of solutions of the forced periodic nonlinear Schrödinger equation. *Nonlinearity*, 23:2183, 2010.

93. Y. Sivan, G. Fibich, B. Ilan, and M.I. Weinstein. Qualitative and quantitative analysis of stability and instability dynamics of positive lattice solitons. *Phys. Rev. E*, 78:046602, 2008.

94. A. Soffer and M. I. Weinstein. Nonautonomous Hamiltonians. *J. Stat. Phys.*, 93, 1998.

95. A. Soffer and M. I. Weinstein. Time dependent resonance theory. *Geom. Func. Anal.*, 8, 1998.

96. A. Soffer and M.I. Weinstein. Ionization and scattering for short lived potentials. *Lett. Math. Phys.*, 48, 1999.

97. A. Soffer and M.I. Weinstein. Resonances, radiation damping, and instability of Hamiltonian nonlinear waves. *Inventiones Mathematicae*, 136:9–74, 1999.

98. A. Soffer and M.I. Weinstein. Selection of the ground state for nonlinear Schrödinger equations. *Reviews in Mathematical Physics*, 16(8):977–1071, 2004.

99. A. Soffer and M.I. Weinstein. Theory of nonlinear dispersive waves and selection of the ground state. *Phys. Rev. Lett.*, 95:213905, 2005.

100. C. Sparber. Effective mass theorems for NLS equations. *SIAM J. Appl. Math.*, 66:820–842, 2006.

101. W.A. Strauss. Existence of solitary waves in higher dimensions. *Comm. Math. Phys.*, 55:149–162, 1979.

102. C. Sulem and P.L. Sulem. *The Nonlinear Schrödinger Equation: Self-Focusing and Wave Collapse*, volume 139 of *Series in Mathematical Science*. Springer-Verlag, 1999.

103. T. Tao. *Nonlinear Dispersive Equations: Local and Global Analysis*. Number 106 in CBMS Regional Conference Series. AMS, 2006.

104. T. Tao. A global compact attractor for high-dimensional defocusing non-linear Schrödinger equations with potential. *Dyn. PDE*, 5:101–116, 2008.

105. T. Tao. Why are solitons stable? *Bulletin and the AMS*, 46(1):1–33, 2009.

106. T.-P. Tsai and H.-T. Yau. Asymptotic dynamics of nonlinear Schrödinger equations: resonance-dominated and dispersion-dominated solutions. *Comm. Pure Appl. Math.*, 55(2): 153–216, 2002.

107. T.-P. Tsai and H.-T. Yau. Classification of asymptotic profiles for nonlinear Schrödinger equations with small initial data. *Adv. Theor. Math. Phys.*, 6(1):107–139, 2002.

108. M. Vakhitov and A. Kolokolov. Stationary solutions of the wave equation in a medium with nonlinearity saturation. *Radiophys. Quant. Elec.*, 16:783, 1973.

109. M. I. Weinstein. Resonance problems in photonics. In D-Y Hsieh, M Zhang, and W Sun, editors, *Frontiers of Applied Mathematics- Proceedings of the 2nd International Symposium*. World Scientific, 2007.

110. M.I. Weinstein. Nonlinear Schrödinger equations and sharp interpolation estimates. *Comm. Math. Phys.*, 87:567–576, 1983.

111. M.I. Weinstein. Modulational stability of ground states of nonlinear Schrödinger equations. *SIAM J. Math. Anal.*, 16:472–490, 1985.

112. M.I. Weinstein. Lyapunov stability of ground states of nonlinear dispersive evolution equations. *Comm. Pure Appl. Math.*, 39:51–68, 1986.

113. V. Weisskopf and E. Wigner. Berechnung der naturlichen Linienbreite auf Grund der Diracschen Lichttheorie. *Z. Phys.*, 63:54–73, 1930.

114. G.B. Whitham. *Linear and Nonlinear Waves*. Wiley-Interscience, 1974.